# Current Topics in Microbiology 205 and Immunology

Springer
Berlin
Heidelberg
New York
Barcelona
Budapest
Hong Kong
London
Milan
Paris
Santa Clara
Singapore
*Tokyo*

# The CD4 Molecule

## Roles in T Lymphocytes and in HIV Disease

Edited by D.R. Littman

With 29 Figures

Springer

Professor DAN R. LITTMAN, M.D., Ph.D.
Howard Hughes Medical Institute
Division of Molecular Pathogenesis
Skirball Institute of Biomolecular Medicine
New York University Medical Center
540 First Avenue
New York, NY 10016
USA

*Cover illustrations: The HIV receptor. A. A ribbon diagram of domain 1 of human CD4 from the standard point of view (e.g., Figure 2 of Brady and Barclay, Chapter 1, this volume). The N-terminus of strand A and the C-terminus of strand G are marked. B. A space-filling model of domain 1 of human CD4 from the point of view of a monomer of gp120 approaching the contact site. In both panels, residues are colored similarly. Residues 86–89 (F-G turn, equivalent to DCR3 of Ig variable domains) are shown in green. Residues 33–62 (C'–C"–D, equivalent to CDR2) are shown in red, with the exception of the following. Residues Lys29, Lys35, Lys46 and Arg59 are shown, with their side-chains, in yellow. Residue Phe43, with its side-chain, is shown in off-white. The remaining residues of the C" ridge (Ser42, Leu44, Thr45, Gly47) are shown in magenta. All other residues are shown in blue.*

*Cover design: Künkel+Lopka, Ilvesheim*

ISBN-13: 978-3-642-79800-9 e-ISBN-13: 978-3-642-79798-9
DOI: 10.1007/978-3-642-79798-9


Library of Congress Catalog Card Number 15-12910
Softcover reprint of the hardcover 1st edition 1996



Typesetting: Thomson Press (India) Ltd, Madras
SPIN: 10485113 27/3020/SPS – 5 4 3 2 1 0 – Printed on acid-free paper

# Preface

During the late 1970's the application of hybridoma technology led to an explosion in the discovery and characterization of proteins expressed at the surface of hematopoietic cells. The understanding of T lymphocyte biology benefited enormously from this advance and from newly developed techniques for obtaining clonal T cells. Application of these methodologies resulted in the identification of the clonally restricted T cell antigen receptors (TCRs) and of a number of other molecules expressed more broadly on T cells. Among these, the CD4 and CD8 glycoproteins stood out because they were differentially expressed on distinct functional subsets of T lymphocytes. Moreover, blocking studies with monoclonal antibodies suggested a functional role for CD4 and CD8 in T cell responses to antigen. Shortly thereafter, it was shown that T helper cells were the primary targets for the human immunodeficiency virus (HIV) and that CD4 serves as the viral receptor on these cells. These findings fueled an intense interest in CD4 during the last decade, in the hope that understanding the molecular nature of the HIV-CD4 interaction could hold the key to controlling AIDS.

The interesting biological function of CD4 was gleaned early on from the observation that its expression is restricted primarily to T helper cells. The finding that the cognate molecule, CD8, is expressed exclusively on cytotoxic T cells and the subsequent recognition that $CD8^+$ cells interact specifically with MHC class I molecules suggested that these glycoproteins may interact directly with different classes of MHC proteins (Swain 1981). Subsequent studies showed that, indeed, CD4 and CD8 bind to regions of MHC class II and class I, respectively, that are distinct from the antigen-binding pockets recognized by the TCR (see Chap.2). Early work with blocking antibodies suggested that CD4 is involved in mediating adhesion, thus strengthening the interaction of "low affinity" TCRs with MHC/antigen (Marrack et al. 1983). The subsequent discovery of the association of the protein tyrosine kinase $p56^{lck}$ with CD4 resulted in a turnaround in the view of CD4 function, with most investigators favoring a signaling role via associated Lck after binding of CD4 to MHC (RUDD 1990). More recent studies suggest that both interpretations are partially correct (KILLEEN et al. 1993).

Cloning of the CD4 gene in the mid-1980's made it possible to examine the structural and functional features of CD4 in depth. These studies have defined the role of CD4 as a co-receptor (JANEWAY 1989) which participates in a complex with the TCR to recognize MHC/antigen and to initiate various signals in developing thymocytes or mature T cells. Genetic manipulation of the CD4 cDNA permitted demonstration of the direct interaction of the gene product with MHC class II. In addition, this allowed large quantities of protein to be produced for crystallization and structural studies. Introduction of mutant forms of CD4 into functional T lymphocytes further permitted dissection of the role of different domains of the molecule in the process of antigen recognition. The most recent studies have exploited gene targeting and transgenic technologies to explore the function of CD4 in the developing thymus.

Although initial attention to CD4 was due to the important role of the molecule in T lymphocyte activation and development, it was the discovery of the essential and complex role of CD4 in the life cycle of HIV that has contributed to much of the recent interest in the structure and function of the glycoprotein. In the mid-to-late 1980's, the terms "T4 cells" and "CD4 cells" made their entry into the popular lexicon, as awareness of the importance of these cells in the progression to AIDS began to reach the lay public. The reviews presented in this volume represent an attempt to provide a comprehensive overview of current research on the functions of CD4 in the normal immune response and in HIV disease. The subject areas have been divided into the following categories: structure and function of the CD4 molecule, particularly as it involves its interaction with MHC class II and the TCR complex (Chaps. 1,2); signaling functions of CD4 and the associated protein tyrosine kinase, $p56^{lck}$ (Chaps. 3,4); the role of the CD4 glycoprotein and of CD4 gene regulation in development (Chap. 5); the cell biology of CD4, including its intracellular interactions with HIV gene products (Chap.6); and the functions of CD4 in HIV entry and pathogenesis (Chaps. 7,8).

Some areas of research on CD4 are covered in greater depth than others, largely because more is known about them. For example, the crystal structures of the two component parts of the extracellular domain of CD4 have been elucidated. Much is also known of the interaction of CD4 with MHC class II and of its requirement for activation via the TCR complex. Recent studies also provide a clearer understanding of the role of CD4 in the development of functional subsets of T lymphocytes. In HIV disease, the interaction of the viral receptor, CD4, with the viral envelope glycoprotein has been worked out in some detail (discussed in Chaps. 1,7), but an in vivo function for CD4 in virus-induced cell death remains a matter for much speculation (Chap. 8).

Several areas of CD4 biology are not covered in this monograph, but their importance should by no means be ignored. One of these, in which our understanding remains sketchy, is the phenomenon known as "negative signaling". This refers to inhibition of TCR-mediated signal transduction when there is prior engagement of CD4 with cross-linking antibodies. This phenomenon has been most clearly described using primary T lymphocytes. For example, cross-linking of CD4 upon binding of HIV envelope glycoprotein has been shown to result in increased apoptosis of human T cells, and has been suggested as a potential mechanism of T cell death in AIDS (see Chap. 8). The signaling processes resulting in signal abrogation or induction of apoptotic cell death have not been elucidated, but one study has suggested that sequestration of $p56^{lck}$ by CD4 from the TCR complex may contribute to this phenomenon (Haughn et al. 1992). There is also evidence that cross-linking of CD4 upon HIV binding results in NFkB translocation to the nucleus, where it would enhance viral gene expression (Benkirane et al. 1994). This result suggests a signaling role for CD4 that is independent of the TCR. However, this area is quite controversial, since another study provides evidence for an inhibitory rather than a stimulatory role for the CD4 cytoplasmic domain in viral replication (Tremblay et al. 1994).

Another area of interest, where a mechanistic understanding is lacking, is the phenomenon of T cell phenotypic switching observed upon treatment of animals with anti-CD4 antibodies. For example, in response to injection of *Leishmania major*, Balb/c mice mount a response that is dominated by antigen-specific $T_h2$ cells rather than protective $T_h1$ cells, and the animals succumb to the infection. If anti-CD4 is injected at the time of infection, a dominant Th1 response is mounted, and the animals are protected (Sadick et al. 1987). Treatment with nondepleting anti-CD4 antibodies has also been shown to result in immunological tolerance to various antigens that are administered concurrently. For example, some strains of mice will accept allogeneic skin grafts when anti-CD4 is administered for a brief period of time after grafting. The animals will then accept subsequent grafts from the same allogeneic source, but will reject third party skin grafts. Most remarkably, $CD4^+$ cells from the tolerized mice can transfer skin graft tolerance to naive recipient mice that never received anti-CD4 treatment (QIN et al. 1993). In these transfers, the host T cells assume the tolerizing function, suggesting that the antigen specific tolerizing T cells, effectively suppressor cells, may shift the host T helper cell response from an inflammatory response to one that keeps the inflammatory T cells in check. The mechanisms through which anti-CD4 effects these apparent switches in T cell function in vivo are not understood. Nevertheless, a large number of studies using anti-CD4 therapy have been initiated. In mouse model systems for autoimmunity, such as NZB/NZW mice which develop a lupus-like disease, anti-CD4 has been shown to be quite effective (WOFSY and CARTERON 1990). Based on these successes in the mouse, several trials of anti-CD4 therapy in human diseases, such as rheumatoid arthritis and psoriasis, are being conducted.

In the pathogenesis of AIDS, the CD4 molecule plays key roles, not only as receptor for HIV, but also as a target of interactions with the HIV gene products Nef and Vpu and, potentially, as a contributor to aberrant T cell activation resulting in premature T cell death. In humans, CD4 is expressed not only on T helper cells, but also on macrophage lineage cells, including microglial cells in the brain. Use of soluble forms of CD4 in vitro can effectively block infection with laboratory strains of HIV. However, primary strains of HIV are relatively resistant to this treatment, in large part due to low affinity of soluble CD4 for the oligomeric envelope glycoproteins of these virions. It is therefore not surprising that trials of soluble CD4 in AIDS patients have to date been unsuccessful. Recent studies suggest that infection of macrophages may be critical early in the course of the disease, raising the possibility that manipulation of CD4 on macrophages, where it is not thought to have an important physiological function, may attenuate viral spread in vivo.

Two of the "nonessential" HIV gene products, Nef and Vpu, have important interactions with CD4. These gene products are dispensable for viral replication in vitro. However, Nef, in the closely related simian immunodeficiency virus (SIV), is essential for induction of disease in infected rhesus macaques. Expression of Nef from HIV and SIV results in down-regulation of CD4 from the cell surface through a mechanism of enhanced endocytosis (see Chap. 6 for a more complete treatment of this subject). Nef, a myristylated membrane-associated protein of 27 kDa, also appears to affect signal transduction via the TCR complex, possibly increasing sensitivity of the cell to antigenic stimulation. It is not yet known if CD4 down-regulation is essential for the in vivo effect of Nef or if the Nef effects on CD4 and on T cell signaling are causally related (LITTMAN 1994). The importance of Vpu in vivo is not yet known (Vpu is absent from SIV, so it is more difficult to study). Vpu is an integral membrane protein that enhances virus particle release and causes degradation of CD4 within the endoplasmic reticulum following an interaction mediated through the cytoplasmic domain of CD4. Degradation of CD4 is dependent on phosphorylation of two serines within Vpu (SCHUBERT and STREBEL 1994). Together, Nef and Vpu thus decrease the level of surface CD4 in HIV-infected cells; this may facilitate release of virus from the cell surface (by decreasing the possibility of virus binding to the cell from which it is released) or the enhanced activation of the T cell through indirect mechanisms, resulting in enhancement of viral transcription and production of infectious particles. Interruption of the interaction of the HIV gene products with CD4 may therefore have practical consequences in the control of HIV disease.

From this brief introduction, it should be clear that a comprehensive understanding of the interactions of CD4 with a number of extracellular and intracellular molecules is likely to contribute to the development of practical approaches for controlling normal and abnormal immune responses and the replication of HIV and its pathogenic consequences.

New York, USA DAN R. LITTMAN

## References

Benkirane M, Jeang K T, and Devaux C (1994) The cytoplasmic domain of CD4 plays a critical role during the early stages of HIV infection in T-cells. EMBO J 13: 5559–5569

Haughn L, Gratton S, Caron L, Sekaly R P, Veillette A, Julius M (1992) Association of tyrosine kinase p56lck with CD4 inhibits the induction of growth through the alpha beta T-cell receptor. Nature 358: 328–331

Janeway C A Jr (1989) The role of CD4 in T-cell activation: accessory molecule or co-receptor? Immunol Today 10: 234–238

Killeen N, Davis C B, Chu K, Crooks M E, Sawada S, Scarborough J D, Boyd K A, Stuart S G, Xu H, Littman D R (1993) CD4 function in thymocyte differentiation and T cell activation. Phil Trans Royal Soc London 342: 25–34

Littman D R (1994) Immunodeficiency viruses. Not enough sans Nef. Curr Biol 4: 618–620

Marrack P, Endres R, Shimonkevitz R, Zlotnik A, Dialynas D, Fitch F, Kappler J (1983) The major histocompatibility complex-restricted antigen receptor on T cells. II. Role of the L3T4 product. J Exp Med 158: 1077–1091

Qin S, Cobbold S P, Pope H, Elliott J, Kioussis D, Davies J, Waldmann H (1993) "Infectious" transplantation tolerance. Science 259: 974–977

Rudd C E (1990) CD4, CD8 and the TCR-CD3 complex: a novel class of protein-tyrosine kinase receptor. Immunol Today 11: 400–406

Sadick M D, Heinzel F P, Shigekane V M, Fisher W L, Locksley R M (1987) Cellular and humoral immunity to Leishmania major in genetically susceptible mice after in vivo depletion of L3T4 + T cells. J Immunol 139: 1303–1309

Schubert U, Strebel K (1994) Differential activities of the human immunodeficiency virus type 1-encoded Vpu protein are regulated by phosphorylation and occur in different cellular compartments. J Virol 68: 2260–2271

Swain S L (1981) Significance of Lyt phenotypes: Lyt2 antibodies block activities of T cells that recognize class 1 major histocompatibility complex antigens regardless of their function. Proc Nat Acad Sci USA 78: 7101–7105

Tremblay M Meloche S, Gratton S, Wainberg M A, Sekaly R P (1994) Association of $p56^{lck}$ with the cytoplasmic domain of CD4 modulates HIV-1 expression. EMBO J 13: 774–783

Wofsy D, Carteron N L (1990) CD4 antibody therapy in systemic lupus erythematosus. Sem Immunol 2: 419–425

# List of Contributors

(Their addresses can be found at the beginning of their respective chapters.)

# List of Contents

# The Structure of CD4

R.L. Brady[1] and A.N. Barclay[2]



## 1 Identification and History of CD4

The CD4 antigen was first identified by the W3/25 monoclonal antibody (mAb) in the rat and indeed was one of the first lymphocyte cell surface antigens to be defined by monoclonal antibodies in any species (Williams et al. 1977). The W3/25 mAb was of particular interest as it was the first marker for the sub-population of T lymphocytes with "helper" activity (White et al.1978). The human homologue of the W3/25 antigen was identified with the mAb T4 (Reinherz et al. 1979) and later termed CD4. The CD (clusters of differentiation) nomenclature was introduced as a simple method to group together mAbs that recognised the same antigen at the surface of human leucocytes (Bernard et al. 1984). This nomenclature is now widely used to describe both the antibody recognising the antigen and the antigen itself. It is also used to describe homologues in species other than humans.

[1]Department of Biochemistry, University of Bristol, University Walk, Bristol BS8 1TD, UK
[2]MRC Cellular Immunology Unit, Sir William Dunn School of Pathology, University of Oxford, South Parks Road, Oxford OX1 3RE, UK

The CD4 antigen rapidly became of considerable interest as the W3/25 mAb was used to show the first effect of a monoclonal antibody on a functional assay in vitro, the inhibition of a mixed lymphocyte reaction (WEBB et al. 1979). It was also effective in vivo in inhibiting an autoimmune disease model, experimental autoimmune encephalomyelitis (BROSTOFF and MASON 1984). CD4 was used extensively as a marker for helper T cells and then correlated with restriction of T cells to MHC class II antigens (reviewed in LITTMAN 1987). CD4 is a membrane glycoprotein of Mr 55 000 which plays a central role in the initiation of T cell responses by interacting with MHC class II antigens on antigen presenting cells that present peptides to the antigen specific receptor on T cells. The effects are transmitted to the cytoplasm via a motif in the cytoplasmic domain with $p56^{lck}$ a tyrosine kinase involved in the activation of T lymphocytes (RUDD et al. 1988: VEILLETTE et al. 1990).

CD4 was shown to be the primary receptor for the human immunodeficiency virus (HIV) that causes AIDS (DALGLEISH et al. 1984; KLATZMANN et al. 1984). Subsequently this interaction of CD4 with the gp120 protein HIV has been of major interest with a view to understanding the pathology of HIV infection. In this review we discuss the determination of the structure of the extracellular part of CD4 and its implications for the function and evolution of CD4.

## 2 Sequence of CD4 and Significance for the Immunoglobulin Superfamily

The amino acid sequence of CD4 was of particular interest and was determined at the cDNA level by a transfection protocol whereby human genomic DNA was transfected into a mouse cell line and a CD4 positive transformant isolated. Human cDNA prepared from this cell line was enriched by subtractive hybridisation and used to isolate CD4 cDNA from human T lymphocyte cDNA libraries (MADDON et al. 1985). The predicted amino acid sequence of CD4 indicated it had a single transmembrane domain and the extracellular region contained four domains as illustrated in Fig. 1. Domain 1 showed clear amino acid sequence similarities to immunoglobulin superfamily (IgSF) domains but assignments for domains 2–4 were controversial (CLARK et al. 1987; MADDON et al. 1985, 1987). In particular domain 3 lacked the disulphide bridge between the β-sheets that is normally conserved in IgSF domains. Domain 2 was predicted to have an unusual disulphide bridge, and domains 2 and 4 to be truncated (CLARK et al. 1987; WILLIAMS et al. 1989). As discussed below these predictions have been confirmed by the structural studies. The fact that these assignments based on sequence analysis were controversial but were correct gives considerable weight to the usefulness of assignment of IgSF domains by analysis of their amino acid sequences (WILLIAMS and BARCLAY 1988; WILLIAMS et al. 1989).

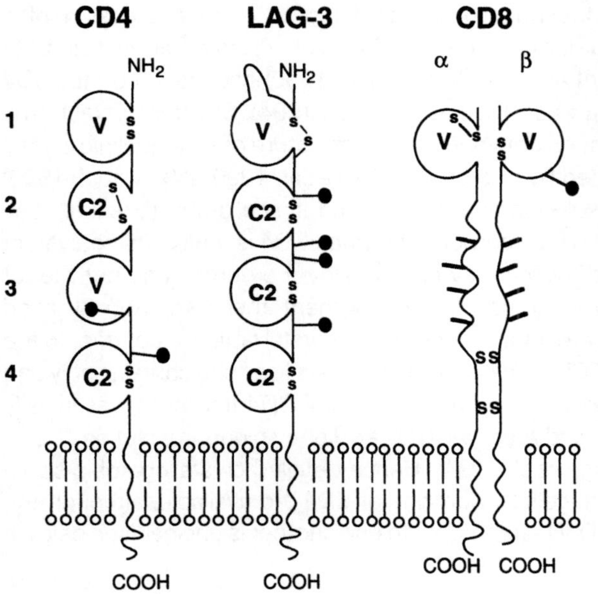


**Fig. 1.** Diagram of human CD4 and comparison with other co-receptors, the related antigen LAG-3 and also CD8. Immunoglobulin superfamily (IgSF) domains are indicated by the *circles* with S-S to indicate the disulphide bridges between the β-sheets of the IgSF domains (CD8 and CD4 domains 2 have unusual disulphide bridges within one β-sheet). The protrusion in domain 1 of LAG-3 represents extra sequence found in the middle of this domain. The domains are numbered from the $NH_2$-terminus

## 3 The Structure of the Extracellular Portion of CD4

Structural analysis of CD4 by X-ray crystallography has centred on the extracellular region expressed by recombinant DNA techniques because of the impracticability of crystallising intact transmembrane glycoproteins. Experience has shown crystallisation of heavily glycosylated and flexible proteins is often difficult. Progress on crystallography of CD4 has relied on studies of fragments of the receptor from two species which differ in their patterns of glycosylation.

The extracellular parts of CD4 (usually residues 1–371) were expressed in both Chinese hamster ovary (CHO) and insect cells, in order to retain both the glycosylation and formation of disulphide bridges. Crystals of this form of human CD4 were obtained readily (Kwong et al. 1990); however, studies from several groups have shown these diffract poorly and are sensitive to radiation exposure. Nevertheless, X-ray data sets to $D_{min}$ = ~5 Å have been collected, although no satisfactory structure solution has been obtained to date (Kwong et al. 1990; Barclay et al. 1993). Although crystals of this form of rat CD4 could not be obtained (possibly reflecting the alternative carbohydrate locations) crystals of a complex of the extracellular domains of rat CD4 bound to an antibody Fab fragment were also obtained (Davis et al. 1990a). These, too, diffract poorly (anisotropically to 3.5–4.5 Å) and are radiation sensitive, limiting collection of X-ray data to 4.5 Å. Removal of either one or both glycosylation sites in rat CD4 did not prevent secretion from CHO cells but diffracting crystals were not obtained (Davis et al. 1990b; Barclay et al. 1993).

Dissection of the CD4 extracellular region led to good expression of a fragment of CD4 containing domains 1 and 2 (DID2) as a single fragment in both *E. coli* and CHO cells (Ryu et al. 1990; Wang et al. 1990) and later the rat D3D4 domains in CHO cells (Brady et al. 1993; Lange et al. 1994). Several crystal forms of the DID2 fragment have been obtained and the structure of a monoclinic crystal form at 2.4 Å was determined by two groups (Ryu et al. 1990; Wang et al. 1990) and later at both higher resolution and in a second space group (Garrett et al. 1993; Ryu et al. 1994). The D1D2 fragment of human CD4 is, unlike its equivalent in rat CD4, not glycosylated, which may explain why the human and not the rat form has crystallised. In the human D3D4 fragment there are two N-linked glycosylation sites, possibly preventing packing in a crystal lattice in contrast to the equivalent fragment from rat CD4 which has been crystallised and contains only one N-linked site (Brady et al. 1993). The structure of this D3D4 fragment was initially reported at 2.8 Å resolution, and more recently at slightly higher resolution (2.6 Å) and lower temperature (Lange et al. 1994). Hence detailed crystallographic structures of each pair of the extracellular domains of CD4 are available, although the orientation of the D1D2 and D3D4 fragments to one another is unclear (see below).

## 4 The Domains of CD4 and the Immunoglobulin Superfamily

The X-ray crystallography structures of the four extracellular domains of CD4 confirmed earlier sequence-based predictions that these domains were members of the IgSF as illustrated in Fig. 2. Domains 1 and 3 are similar to antibody variable region (V-set) domains, while domains 2 and 4 are closer to the size of C-domains but with patches of sequence similar to V-domains and this goup is termed C2-set domains (Williams and Barclay 1988). In particular the correct predictions for domains 2 and 3 were important, as some of the conserved features found in IgSF domain cores, and by which these domains are usually recognised, are absent (Williams and Barclay 1988; Williams et al. 1989). In domains 1 and 3 the nine strands form two β-sheets which can be termed ABED and GFCC′C″ by analogy with antibody V-domains. In antibodies the A strand changes from one sheet to the other half way along its length. In CD4 D1 and D3 the A strand is shorter and is hydrogen-bonded to the B strand in its entirety (although the A strand may be artificially shortened in the D3D4 construct). The fold of D1 is remarkably similar to the REI antibody variable light chain dimer (rms deviations of core residues 2 Å (Ryu et al. 1990; Wang et al. 1990)) and this domain shares all of the core features conserved in antibody domains (disulphide, C strand tryptophan, salt-bridge etc.).

In overall topology D3 has similarities with an antibody V-domain but the construction of its core is different. The intersheet disulphide is absent (replaced by a Phe-Leu pair), the C strand Trp is located further along the C strand and the

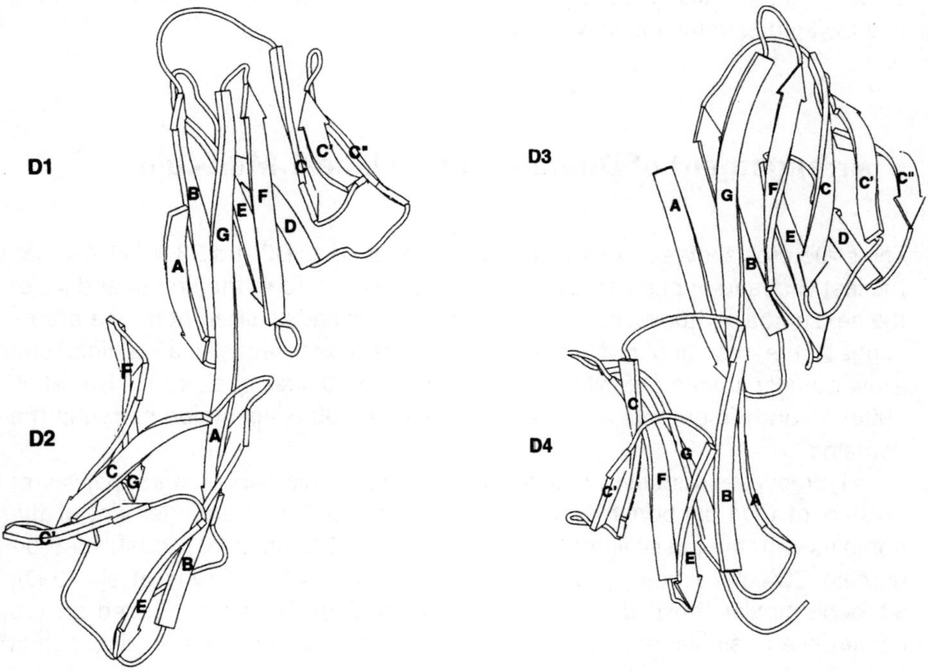


**Fig. 2.** The folding pattern of CD4. The structure of CD4 D1D2 from human is compared to the structure of CD4 D3D4 from rat. The β-sheets are represented by *broad arrows* and the loops between the sheets as *thin lines*. The D1D2 structure is from (Ryu et al. 1990; Wang et al. 1990) and D3D4 from (Brady et al. 1993)

characteristic salt bridge between strands D and F is absent (Lange et al. 1994). Associated with these changes is an alteration in the packing of the β-sheets, which move significantly further apart and "slide" relative to their equivalent positions as observed in antibodies and in D1. These alterations are not solely the result of losing the disulphide, as domain 1 of CD2 is also a V-domain lacking the inter-sheet disulphide but does not show a similar widening (Jones et al. 1992).

The structure shows that D2 and D4 are characterised by a short connecting stand between the C and E strands, termed the C' strand, which hydrogen bonds to strand C rather than E as is observed for strand D in C domains. CD4 D2 has no intersheet disulphide but an unusual intrasheet disulphide connecting the F strand and C strand (the C strand Cys takes the place of the conserved Trp). Again, as in D3, this loss of the intersheet connection is associated with looser packing of the sheets, which slide and are further apart than is normally observed (Brady et al. 1993; Garrett et al. 1993). In D4 the usual disulphide and inter-sheet spacing is observed. A distinctive feature of these two domains of CD4 is the presence of a Trp in the position normally occupied by Tyr in antibodies in strand F (this is the most highly conserved residue in IgSF domains). In D2 and D4 the amine group of the imidazole ring forms a hydrogen bond to the main chain carbonyl of the residue four before the Trp, performing the same function as the hydroxyl group in the usual tyrosine.

## 5 Arrangement of Domains into a Linear Molecule

The structures showed a unique arrangement of D1 to D2 and D3 to D4 in which the last G strand of one domain continues directly to form the first strand (A) of the next domain. Studies of different crystal forms and at different temperatures suggest the resulting rod-like structure behaves essentially as a rigid unit allowing only limited flexibility between domains (LANGE et al. 1994; RYU et al. 1994). In both fragments a similar surface area (900 Å$^2$) is buried between the domains.

Hydrodynamic studies coupled with electron microscopy and analysis of crystals of CD4 support the notion of an extended linear arrangement of the domains with the possibility of flexibility at the D2/D3 junction. Models for four-domain CD4 have been proposed (BRADY et al. 1993; GARRETT et al. 1993); although similar these differ principally at the D2/D3 junction. Based on the occurrence of similar residue types GARRETT et al. (1993) proposes this junction to be rigid and similar to those observed between the D1 and D2 and also between D3 and D4. Based on electron microscopy studies our own model (BRADY et al. 1993) suggests the junction to be more open and flexible which could result in a number of conformations existing. Thus a definitive model accounting for the arrangement of all four extracellular domains of CD4 is not available.

A linear arrangement of the four extracellular domains of CD4 results in a structure about 120 Å in length. If fully extended this arrangement projects the $NH_2$-terminal domains significantly away from the cell surface. This length also corresponds approximately with the expected lengths of the complex of MHC class II interacting with the T cell receptor (TcR) as illustrated in Fig. 3. Although the details of this complex are not known, there is good evidence for much of the surface of CD4 D1 and possibly parts of D2 to be involved directly in binding to the MHC class II protein (see below). Given the approximately linear arrangement of CD4 domains, it seems likely that other regions of CD4 will interact directly with the TcR complex. A pair of protruding loops in D4 have been suggested as likely contact sites for other proteins (BRADY et al. 1993), although this remains to be confirmed.

Many signals are transduced through the cell membrane following ligand binding by dimerisation of the receptor and its associated cytoplasmic tyrosine kinases. Given the association of CD4 with tyrosine kinases it seems possible that dimerisation of CD4 or association of CD4/kinase with another kinase may be key elements in activation of T cells. There have been suggestions that CD4 is itself a dimer (e.g., RYU et al. 1990; LANGEDIJK et al. 1993) although there are few

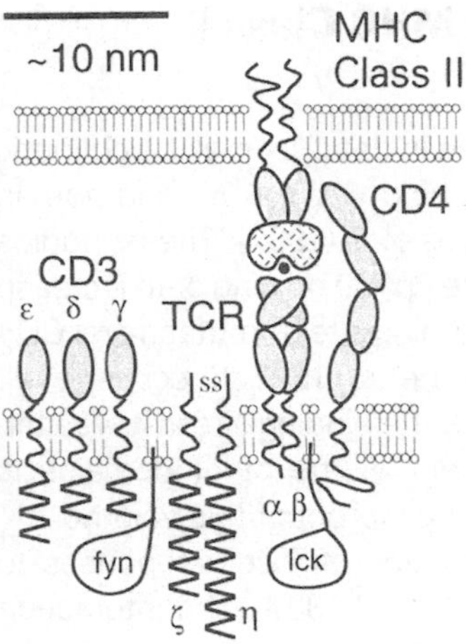


**Fig. 3.** Model for the interaction of MHC class II and the T cell receptor (TCR) together with CD4. The dimensions of MHC class II are from the X-ray crystallography structure of the extracellular region (BROWN et al. 1993), CD4 from the models discussed and TcR based on the dimensions of another protein containing two IgSF domains, CD2 (JONES et al. 1992), but allowing for the additional sequences in the membrane proximal region of the TcR. The IgSF domains are indicated by the *ovals* and the peptide binding domain of MHC class II by *stippling*. The peptide is indicated by the *spot*. Other components indicated are the kinases $p56^{lck}$ and $p60^{fyn}$ and the chains of the CD3 antigen. The diagram indicates the various components but these are not necessarily always associated or in this stoicheiometry

data to support this proposal. For instance the hydrodynamic analysis of recombinant forms of the extracellular part of CD4 showed no propensity to associate even at high concentrations (DAVIS et al. 1990b). D1 differs most noticeably from antibodies in the length of the C'/C" and F/G loops: these are the dimerisation elements in antibody V-domains. In D1 these are foreshortened and one of these (C'/C") forms the basis of the HIV binding site. Not surprisingly, dimerisation of D1 has not been observed. In D3 these loops are of similar length to their equivalents in antibodies; an observation that originally led to the suggestion that CD4 might dimerise using these loops (RYU et al. 1990). However, in the crystal structure they adopt different conformations from those observed in antibodies. A dimeric interaction in crystals of the D3D4 fragment is believed to be the result of crystal packing and of no physiological significance (BRADY et al. 1993). It has been suggested that the location of a conserved glycosylation site on the F strand might act to preclude dimerisation involving the strands used in immunoglobulins (LANGE et al. 1994).

It is possible that association between CD4 molecules occurs through interactions of the transmembrane and cytoplasmic domains or via other proteins and this would not be seen in studies on the recombinant extracellular domains. Studies on the mobility of CD4 on T cell lines using fluorescence photobleaching recovery measurements indicate that about 60% of the CD4 was mobile; however comparison of mobility determinations using Fab and IgG anti-CD4 indicated that the latter were very much slower (PAL et al. 1991). One explanation is that some of the CD4 may be dimeric and this can form multimeric complexes with bivalent antibodies. However one might expect endocytosis of dimers of CD4 to occur with IgG CD4 mAbs whereas in fact cross-linking with anti-Ig is necessary. The X-ray crystal structure of MHC class II suggests that it may be dimeric itself (BROWN et al. 1993). It is probable that productive interactions may be of low affinity and difficult to quantify (see Sect. 6).

## 6 Binding sites on CD4 for HIV-1 and MHC Class II

The binding site on CD4 for HIV-1 has been analysed in detail by making chimeric molecules between mouse CD4, which does not bind HIV gp120, and human CD4, which binds with high affinity ($K_D$~$10^{-9}$; MOEBIUS et al. 1992). These studies have been extended by mutagenesis to either delete gp120 binding or to increase the affinity of the interaction and also to recreate a binding site by mutating rat CD4 residues to the equivalent human residues (MOEBIUS et al. 1992; SHOCKMEL et al. 1992; TSUI et al. 1992). All these studies indicate that the binding site involves the C' C" loops of domain 1. The possible involvement of the other domains is discussed in (RYU et al. 1994) and reviewed in other chapters of this volume.

CD4 is clearly involved in the process of restriction of T cell responses to antigens presented on MHC class II (reviewed in LITTMAN 1987). The interaction between CD4 and MHC class II has been shown between monolayers of cells expressing high levels of MHC class II antigens and purified CD4 (DOYLE and STROMINGER 1987) but this interaction has been difficult to show with purified proteins implying that the interaction is either of very low affinity or that other proteins are involved. Thus sCD4 does not bind to MHC class II expressing cells even when added at high concentrations and does not inhibit MHC class II restricted responses such as the mixed leucocyte interaction even though this is inhibited by antibodies to either the MHC antigens on the stimulators or CD4 on the responders (HUSSEY et al. 1988). However multimeric chimeric CD4 proteins have recently been shown to bind cells expressing high levels of MHC class II implying a very low affinity (WEBER and KARJALAINEN 1993). There are two contrary reports; CAMMAROTA et al. (1992) showed that purified recombinant sCD4 bound soluble MHC class II antigens that had been prepared without transmembrane regions with a $K_D$=3.2 x $10^{-6}M$. In addition they showed an interaction between CD4 and either isolated MHC class II β-chain or peptides from the class II antigen. Comparable results have been obtained by NAG et al. (1993) who showed binding of purified MHC class II heterodimer and isolated β-chain, but not α-chain, to cells expressing high levels of CD4. The $K_D$ was not measured but saturating binding was obtained at around 10n*M* which points to a high affinity. These values for a relatively high affinity are at odds with the inability to obtain binding of sCD4 to cells and with the data of WEBER and KARJALAINEN (1993). If the $K_D$ value of CAMMAROTA et al. is correct, one would not expect to see any binding at the concentrations used by NAG et al. (1993). One explanation may be that there are aggregates in the preparation of recombinant MHC class II antigens giving rise to overestimates of the affinity. Even small amounts of aggregated material can give very misleading results as shown recently in studies of the interaction between CD2 and CD48 (VAN DER MERWE et al. 1993). Aggregates are particularly likely to be present in the study of NAG et al. (1993) as the MHC class II antigens were purified from cells with detergent but the binding is done in the absence of detergent, a procedure that is known to give rise to micelles in other cell surface proteins (KUCHEL et al. 1978). If activation through the TcR involves a trimolecular complex between MHC class II, CD4 and TcR itself, then one may anticipate that the CD4-class II affinity must be low if antigen specificity of the trimolecular complex is to be preserved. In conclusion CD4 is most likely to interact with the β-chain of MHC class II in a comparable manner to the interaction of CD8 and MHC class I but with a very low affinity.

Mutagenesis studies have on the whole indicated a larger part of the surface of CD4 is involved in MHC class II binding than for HIV although there have been some differences depending on assays used. This is discussed in detail elsewhere (MOEBIUS et al. 1992; GARRETT et al. 1993; HOULGATTE et al. 1994; RYU et al. 1994; VIGNALI et al. 1993 and this volume).

## 7 Analysis of CD4 from Other Species

The amino acid sequence of the extracellular portion of CD4 is relatively poorly conserved across species as shown in Fig. 4 for CD4 sequences from human mouse, rat, rabbit, cat and dog. For example there is only about 53% sequence identity between rat and human CD4 sequences but this is not atypical for the extracellular regions of leucocyte surface proteins. The most striking difference is in the sequence of cat CD4 which has an insertion of 16 residues, rich in Gly and Ser, in the region predicted to be between strands A and B in domain 2. This sequence is likely to form a flexible loop.

In comparisons between IgSF proteins it is usually the inpointing residues of the β strands that show the most similarity with most variability in the loops. This is generally true for comparisons between the CD4 sequences although there are some notable exceptions. There is highly conserved region at the beginning of domain 2 which includes most of the A/B loop and much of the B strand. This loop forms many of the proposed contacts with domain 3 in the model proposed for CD4 above. The loop between strands A and B in domain 3 is conserved with the sequence EGE (EGD in dog). The first of the Glu residues is involved in a salt bridge with Lys-290 in the A strand of D4, which appears important in maintaining the relative conformations of the domains as discussed above. This Lys (or Arg) is conserved in all species sequenced. The Gly-192 places the second Glu or Asp (Glu-193 in rat) in a position to form a charged hydrogen bond with Ser-190, which in turn is hydrogen bonded to Ser-355 at the beginning of the G-strand in domain 4. This latter Ser is totally conserved, and the former is either Ser or Lys, preserving either a charged hydrogen bond or salt bridge with Glu-193. This conserved network of salt bridges and hydrogen bonds in the A/B loop may be essential to maintain the relative orientation of D3 with respect to D4. The F strands in domains 3 and 4 are particularly well conserved.

The unusual disulphide bond in CD4 domain 2 bridging strands C and F (rather than the normal B and F) is not conserved in cat CD4 which lacks the Cys in strand C. However cat CD4 contains an extra Cys in the normally conserved Tyr/Trp

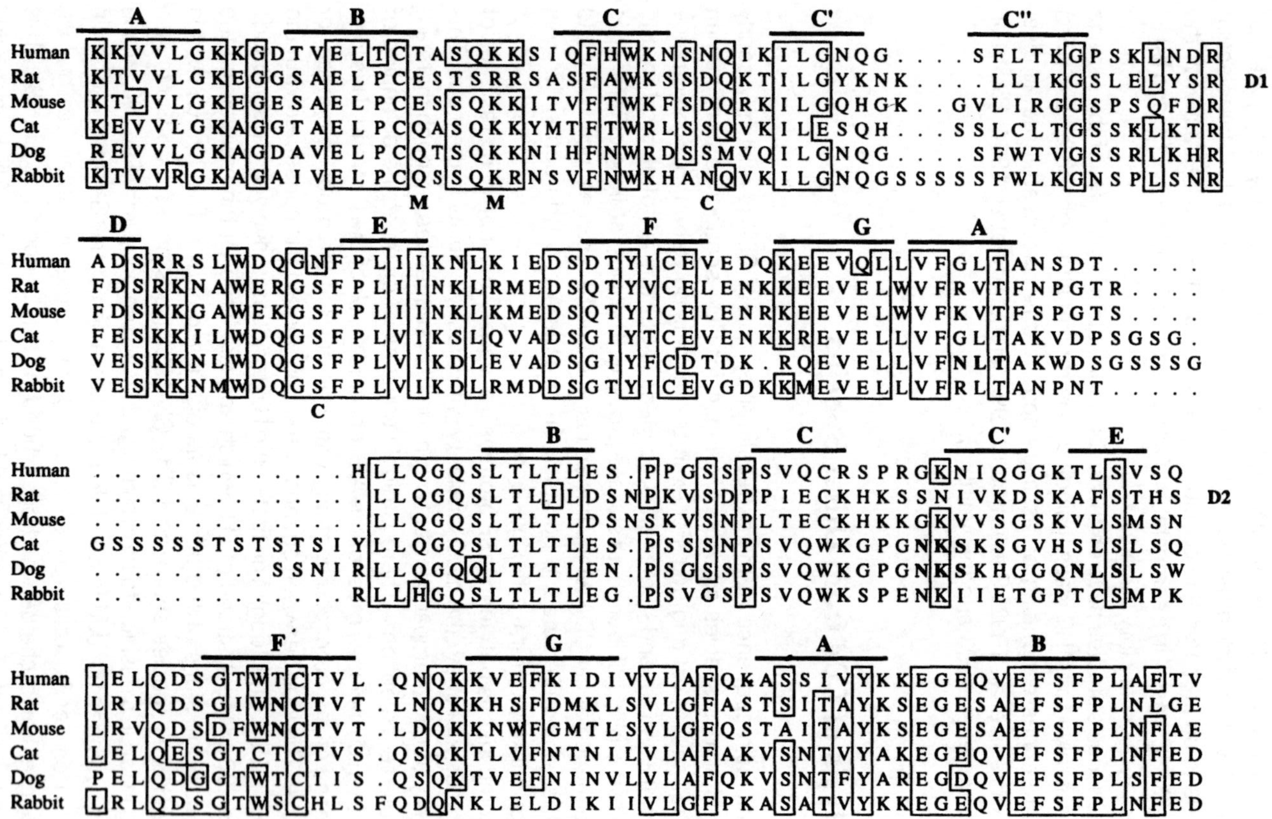
A B C C' C''
Human K K V V L G K K G D T V E L T C T A S Q K K S I Q F H W K N S N Q I K I L G N Q G . . . . S F L T K G P S K L N D R
Rat K T V V L G K E G G S A E L P C E S T S R R S A S F A W K S S D Q K T I L G Y K N K . . . . L L I K G S L E L Y S R
Mouse K T L V L G K E G E S A E L P C E S S Q K K I T V F T W K F S D Q R K I L G Q H G K . . G V L I R G G S P S Q F D R
Cat K E V V L G K A G G T A E L P C Q A S Q K K Y M T F T W R L S S Q V K I L E S Q H . . . S S L C L T G S S K L K T R
Dog R E V V L G K A G D A V E L P C Q T S Q K K N I H F N W R D S S M V Q I L G N Q G . . . . S F W T V G S S R L K H R
Rabbit K T V V R G K A G A I V E L P C Q S S Q K R N S V F N W K H A N Q V K I L G N Q G S S S S S F W L K G N S P L S N R
D1
M M C
D E F G A
Human A D S R R S L W D Q G N F P L I I K N L K I E D S D T Y I C E V E D Q K E E V Q L L V F G L T A N S D T . . . . .
Rat F D S R K N A W E R G S F P L I I N K L R M E D S Q T Y V C E L E N K K E E V E L W V F R V T F N P G T R . . . .
Mouse F D S K K G A W E K G S F P L I I N K L K M E D S Q T Y I C E L E N R K E E V E L W V F K V T F S P G T S . . . .
Cat F E S K K I L W D Q G S F P L V I K S L Q V A D S G I Y T C E V E N K K R E V E L L V F G L T A K V D P S G S G .
Dog V E S K K N L W D Q G S F P L V I K D L E V A D S G I Y F C D T D K . R Q E V E L L V F N L T A K W D S G S S S G
Rabbit V E S K K N M W D Q G S F P L V I K D L R M D D S G T Y I C E V G D K K M E V E L L V F R L T A N P N T . . . . .
C
B C C' E
Human . . . . . . . . . . . . . . H L L Q G Q S L T L T L E S . P P G S S P S V Q C R S P R G K N I Q G G K T L S V S Q
Rat . . . . . . . . . . . . . . . L L Q G Q S L T L I L D S N P K V S D P P I E C K H K S S N I V K D S K A F S T H S
D2
Mouse . . . . . . . . . . . . . . . L L Q G Q S L T L T L D S N S K V S N P L T E C K H K K G K V V S G S K V L S M S N
Cat G S S S S S T S T S T S I Y L L Q G Q S L T L T L E S . P S S S N P S V Q W K G P G N K S K S G V H S L S L S Q
Dog . . . . . . . . . S S N I R L L Q G Q Q L T L T L E N . P S G S S P S V Q W K G P G N K S K H G G Q N L S L S W
Rabbit . . . . . . . . . . . . . . R L L H G Q S L T L T L E G . P S V G S P S V Q W K S P E N K I I E T G P T C S M P K
F' G A B
Human L E L Q D S G T W T C T V L . Q N Q K K V E F K I D I V V L A F Q K A S S I V Y K K E G E Q V E F S F P L A F T V
Rat L R I Q D S G I W N C T V T . L N Q K K H S F D M K L S V L G F A S T S I T A Y K S E G E S A E F S F P L N L G E
Mouse L R V Q D S D F W N C T V T . L D Q K K N W F G M T L S V L G F Q S T A I T A Y K S E G E S A E F S F P L N F A E
Cat L E L Q E S G T C T C T V S . Q S Q K T L V F N T N I L V L A F A K V S N T V Y A K E G E Q V E F S F P L N F E D
Dog P E L Q D G G T W T C I I S . Q S Q K T V E F N I N V L V L A F Q K V S N T F Y A R E G D Q V E F S F P L S F E D
Rabbit L R L Q D S G T W S C H L S F Q D Q N K L E L D I K I I V L G F P K A S A T V Y K K E G E Q V E F S F P L N F E D

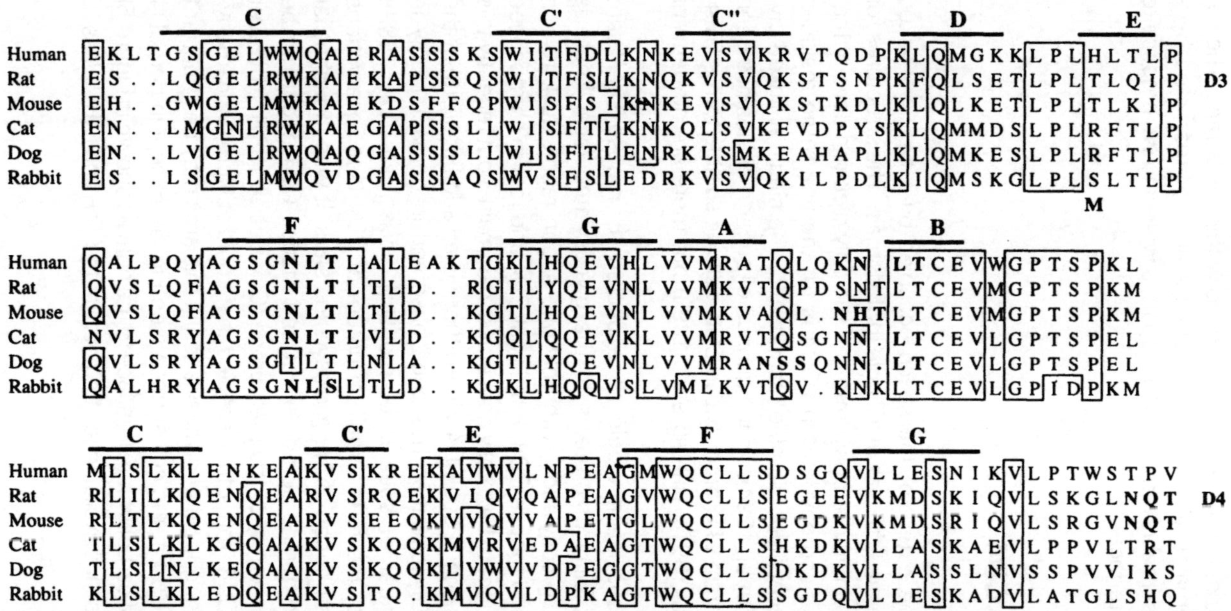


**Fig. 4.** Alignment of the sequences of the extracellular regions of CD4 from human, rat, mouse, cat, dog and rabbit. Data are from the SWISSPROT database. Sequences for CD4 of chimpanzee and various monkeys are not included as these are very similar to human, with 98% and 90% identities respectively. However chimpanzee CD4 has two potential N-glycosylation sites in domain 1 and some monkeys have two different sites also in domain 1 (FOMSGAARD et al. 1992). The positions of these N-linked sites are indicated by the letters *C* and *M* in *bold* beneath the line-ups for chimpanzee and monkey CD4 respectively. Identities in five or more sequences are *boxed* and the positions of the β-strands indicated by the *broad lines* above the sequences. Potential N-glycosylation sequons are shown in *bold*

position in strand F which could form a disulphide bond within this strand. The net effect on the domain remains the same: the sheets are not restricted by a disulphide bridge and presumably remain further apart, a widening we propose to be linked to the D2/D3 interface.

## 8 Glycosylation of CD4

It is not often realised the extent of the space occupied by a typical N-linked carbohydrate but it approaches the size of a single IgSF domain. The structures for human D1D2 do not contain any carbohydrate and the only electron density that can be readily assigned to carbohydrate in D3D4 is for the first three residues (N-acetylglucosamine-*N*-acetylglucosamine-mannose) of the single N-linked glycosylation present (LANGE et al. 1994). The sequence of sugar residues has been determined for rat and human CD4 produced in CHO cells as soluble forms (YUEN et al. 1990; SPELLMAN et al. 1991; ASHFORD et al. 1993) but there are no data for membrane CD4.

The bulk of the glycosylation would be expected to impose significant limitations on the accessibility of proteins to the glycosylated face. There is species variation in CD4 glycosylation with only one site in rabbit CD4, two in human CD4, three in rat and mouse CD4, four in cat CD4 and four in chimpanzee (Fig. 4). One site was thought to be totally conserved in domain 3 but recently this has been shown to be lacking in dog CD4 where there is an additional site at the beginning of domain 4. Glycosylation has been observed in domain 1 only in chimpanzees and a number of species of monkey (FOMSGAARD et al. 1992). Figure 5 shows models of the four-domain structure of CD4 (see reservations above) to illustrate the distribution of the glycosylation sites. It was thought that the site near the transmembrane portion of rat and mouse CD4 might not be glycosylated due to the proximity of the membrane (CLARK et al. 1987). However the presence of sites on CD4 from other species that could still project carbohydrate in a similar direction argues that glycosylation in this area may be important for the orientation of the CD4 at the cell surface (see Fig. 5). Except in dog CD4 one face tends to be free of carbohydrate (the face to the right as shown in Figure 5) perhaps to keep part of this face available to interact with the TcR and MHC class II.

## 9 Evolution of CD4

The similarity between the structures of pairs of domains, that is D1D2 and D3D4 with their unique linkage between the domains, i.e., the continuous β strand between domains 1 (G strand) and 2 (A strand) and a similar strand between domains 3 and 4, is compatible with evolution via a two-domain precursor. This

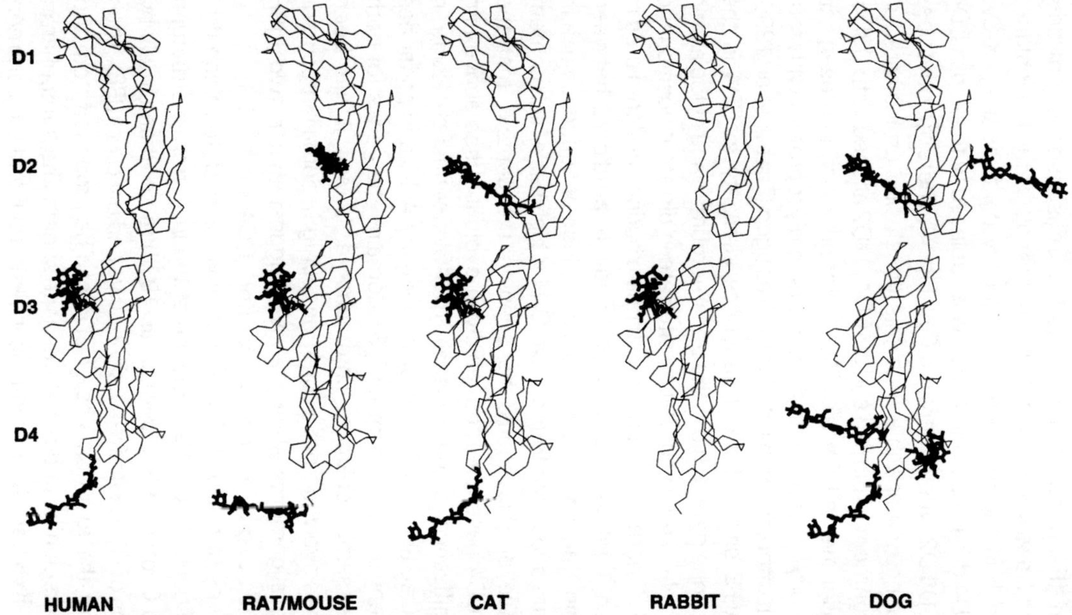


**Fig. 5.** Models of CD4 indicating the different positions of N-linked glycosylation predicted for CD4 from various species. The orientation of D1D2 to D3D4 is controversial with only modelling data available based on the structures for human D1D2 and rat D3D4 (see above). The N-linked structures illustrated are modelled on those observed in analysis of recombinant soluble CD4 and could differ significantly from the membrane bound form. CD4 from various species of monkeys all show conservation of the two sites illustrated for human CD4 with additional sites in domain 1 (see Fig. 4)

had been suggested on the basis of some patches of sequence similarity between parts of domains 2 and 4 and 1 and 3 (WILLIAMS et al. 1989). For instance the sequence in β-strand F in the IgSF almost invariably contains a Tyr-X-Cys sequence. However in CD4 domains 2 and 4 the Tyr is replaced with a Trp in CD4 in all species except cat CD4 D2 where there is a Cys (Fig. 4). There are other less striking patches of sequence compatible with this conclusion as indicated in Fig. 6 which shows an alignment of human and rat CD4 D1D2 aligned with D3D4. This figure also shows alignments with the LAG-3 protein which is present on activated human T lymphocytes and is a transmembrane glycoprotein with four IgSF domains. Like CD4, it interacts with MHC class II antigen (TRIEBEL et al. 1990; BAIXERAS et al. 1992). LAG-3 shows limited sequence similarity to CD4 in the extracellular domains but none detectable in the transmembrane and cytoplasmic regions. All four predicted N-linked glycosylation sites are in different positions to those in CD4. Unlike in CD4, normal intersheet disulphide bridges are predicted in LAG-3 apart from domain 1 where a disulphide bridge is predicted between β-strands B and G. Domain 1 also has the unusual extra loop of sequence rich in Pro. It differs from the extra sequence found in cat CD4 which has a high content of Ser (see Fig. 4). However this should be seen in the context that the extracellular domains of CD4 have diverged rapidly in evolution (see above and Fig. 4). Of particular significance is the unusual Trp-X-Cys sequence found in domains 2 and 4 of CD4 (see above) which is also found in LAG-3 and so far has not to our knowledge been found in any other IgSF domain. The genes for both antigens are both in the same region of chromosome 12 and it has been suggested that these genes evolved from a common four IgSF domain precursor (TRIEBEL et al. 1990). The alignments shown above suggest that a two IgSF domain structure duplicated to give a precursor for both CD4 and LAG-3.

Now that the structures of many IgSF domains have been determined it is possible to use these to speculate on the evolution of the IgSF from a structural point of view (HARPAZ and CHOTHIA 1994; LANGE et al. 1994) rather than simply by sequence analysis in terms of C1, C2 and V-sets (WILLIAMS and BARCLAY 1988). Thus a hierarchy of evolution for the four domains of CD4 can be proposed from the analysis of the conserved disulphide bridge between the sheets, the tryptophan in the C-strand, the tyrosine two before the cysteine in β-strand F and the common salt bridge between the ends of strands D and F (see Fig. 4); these structural features are all highly conserved in Ig domains and hence benchmarks from which to measure divergence (LANGE et al. 1994). The core of D1 retains all of these features, and based on these criteria has diverged least. D3 by contrast retains only the C strand Trp, and that in a non-standard position, and would appear the most divergent of the four domains (and indeed the most divergent structure determined to date of a commonly agreed IgSF domain). D4, retaining the

**Fig. 6.** Alignment of sequences for CD4 D1D2 vs D3D4 and corresponding sequences of the LAG-3 protein. Residues identical in four or more sequences are *boxed*. The approximate position of the β-strands is indicated by *bars* above the sequences and is based on the human CD4 D1D2 structure

```
                A                 B                          C                                      C'
hCD4 D1D2    K K V V L G K K G D T V E L T C T A S . . . . . . . Q K K S I Q F H W K N . . . . . . . . . . . . . S N Q I K I L G N Q .
rCD4 D1D2    K T V V L G K E G G S A E L P C E S T . . . . . . . S R R S A S F A W K S . . . . . . . . . . . . . S D Q K T I L G Y K .
hLAG3 D1D2   V P V V W A Q E G A P A Q L P C S P T I P L Q D L S L L R R A G V T W Q H G   27 residues insert   P R P R R Y T V L S V G
hCD4 D3D4    S S I V Y K K E G E Q V E F S F P L A F T V E . . K L T G S G E L W W Q A E R     (See below)      A S S S K S W I T F D L K
rCD4 D3D4    S I T A Y K S E G E S A E F S F P L N L G E E . . . S L Q G E L R W K A E K . . . . . . . . A P S S Q S W I T F S L K
hLAG3 D3D4   P L T V Y A G A G S R V G L P C R L P A G V . . . G T R S F L T A K W T P . . . . . . . . . . P G G . . . . . . . . .

                  C''              D                E                       F                 G
hCD4 D1D2    G S F L T K G P S K L N D R A D S R R S L W D Q G N F P L I I K N L K I E D S D T Y I C E V E D Q K E E V Q L
rCD4 D1D2    N K L L I K G S L E L Y S R F D S R K N A W E R G S F P L I I N K L R M E D S Q T Y V C E L E N K K E E V E L
hLAG3 D1D2   P G G L R S G R L P L Q P R V Q L D E R G R Q R G D F S L W L R P A R R A D A G E Y R A A V H L R D R A L S C
hCD4 D3D4    N K E V S V K R V T Q D P K L Q M G K K L . . P L H L T L P Q A L P Q Y A G S G N L T L A L E A K T G K L H Q
rCD4 D3D4    N Q K V S V Q K S T S N P K F Q L S E T L . . P L T L Q I P Q V S L Q F A G S G N L T L T L D . . R G I L Y Q
hLAG3 D3D4   . . . . . . G P D L L V T G D N . . . . . . . . G D F T L R L E D V S Q A Q A G T Y T C H I H L Q E Q Q L N A

                A                                    B                 C
hCD4 D1D2    L V F G L . . . . . . T A N S D T H L L Q G Q S L T L T L E S . . P P G S S P S V Q C R S P R G K . . . . . .
rCD4 D1D2    W V F R V . . . . . . T F N P G T R L L Q G Q S L T L I L D S N P K . V S D P P I E C K H K S S . . . . . . .
hLAG3 D1D2   R L R L R L G Q A S M T A S P P G S L R A S D W V I L N C S F S R P . D R P A S V H W F R N R G Q G R V P V R
hCD4 D3D4    E V N L V . . . . . . . . . . V M R A T Q L Q K N . L T C E V W G P . . T S P K L M L S L K L E . . . . . .
rCD4 D3D4    E V N L V . . . . . . . . . . V M K V T Q P D S N T L T C E V M G P T S P K M R L I L K Q E . . . . . . . .
hLAG3 D3D4   T V T L A . . . . I I T V T P K S F G S P G S L G K L L C E V T . P V S G Q E R F V W S S L D T P . . . . . .

                 C'            E                     F                    G
hCD4 D1D2    . . N I Q G G K . . T L S V S Q L E L Q D S G T W T C T V L Q N Q K K V E F K . I D I V V L A F Q K A
rCD4 D1D2    . N I V K D S K . . A F S T H S L R I Q D S G I W N C T V T L N Q K K H S F D . M K L S V L G F A S T
hLAG3 D1D2   E S P H H H L A E S F L F L P Q V S P M D S G P W G C I L T Y R D G F N V S I M Y N L T V L G L E P P
hCD4 D3D4    N K E A K V S K . . R E K A V W V L N P E A G M W Q C L L S D S G Q V L L E S . . N I K V L P T W S T
rCD4 D3D4    N Q E A R V S R . . Q E K V I Q V Q A P E A G V W Q C L L S E G E E V K M D S . . K I Q V L S K G L N
hLAG3 D3D4   . S Q R S F S G P . W L . E A Q E A Q L L S Q P W Q C Q L Y Q G E R L L G A A . V Y F T E L S S P G A

Insert in LAG-3     Q P D S G P P A A A P G H P L A P G P H P A A P S S W
```

standard disulphide but losing the C-strand Trp, is closer to the Ig fold than D2 with its non-strandard disulphide, and therefore presumably closer to a C2 set precursor. Divergence of D2 and D3 may be related to an evolutionary requirement for increased widths of these domains, possibly to provide an interface between them. This is also consistent with divergence of these domains after gene duplication to provide the initial four-domain precursor.

## 10 Concluding Remarks

The structure of CD4 has confirmed that it contains four IgSF domains in a linear array although these superfamily assignments based on amino acid sequence analysis had been controversial. Regions of the proteins that interact with MHC class II and the HIV-1 have been mapped in detail but the orientation of D1D2 to D3D4 is not known. A detailed understanding of the molecular reactions that lead to T cell activation is still unclear with little known about the kinetics of interactions between the proteins known to be involved. It seems likely that one of the roles of CD4 is to bring $p56^{lck}$ together with the TCR complex. Whilst reactions can occur without this interaction in vitro, e.g., by cross-linking with antibodies, CD4 does have a key role in normal T cell development as shown by a wealth of data ranging from blocking of reactions with antibodies to recent experiments involving the production of mouse strains deficient in CD4 or expressing variants thereof as discussed in detail in other chapters of this volume.

*Acknowledgments.* We are grateful to Marion Brown, Simon Davis, Don Mason and William James and for helpful comments. Leo Brady is supported by a MRC AIDS Senior Fellowship.

## References

Ashford DA, Alafi CD, Gamble VM, Mackay DJ, Rademacher TW, Williams PJ, Dwek RA, Barclay AN, Davis SJ, Somoza C, Ward HA, Williams AF (1993) Site specific glycosylation of recombinant rat and human soluble CD4 variants expressed in Chinese hamster ovary cells. J Biol Chem 268: 3260–3267

Baixeras E, Huard B, Miossec C, Jitsukawa S, Martin M, Hercend T, Auffray C, Triebel F, Piatier TD (1992) Characterization of the lymphocyte activation gene 3-encoded protein. A new ligand for human leukocyte antigen class II antigens. J Exp Med 176: 327–337

Barclay AN, Brady RL, Davis SJ, Lange G (1993) CD4 and the immunoglobulin superfamily. Philos Trans R Soc Lond Biol 342: 7–12

Bernard A, Boumsell L, Dausset J, Milstein C, Schlossman SF (1984) Leucocyte typing. Springer, Berlin Heidelberg New York

Brady RL, Dodson EJ, Dodson GG, Lange G, Davis SJ, Williams AF, Barclay AN (1993) Crystal structure of domains 3 and 4 of rat CD4; relation to the NH2-terminal domains. Science 260: 979–983

Brostoff SW, Mason DW (1984) Experimental allergic encephalomyelitis: successful treatment in vivo with a monoclonal antibody that recognizes T helper cells. J Immunol 133: 1938–1942

Brown JH, Jardetsky TS, Gorga JC, Stern LJ, Urban RG, Strominger JL, Wiley DC (1993) Three-dimensional structure of the human class II histocompatibility antigen HLA-DR1. Nature 364: 33–39
Cammarota G, Scheirie A, Takacs B, Doran DM, Knorr R, Bannwarth W, Guardiola J, Sinigaglia F(1992) Identification of a CD4 binding site on the beta 2 domain of HLA-DR molecules. Nature 356: 799–801
Clark SJ, Jefferies WA, Barclay AN, Gagnon J, Williams AF (1987) Peptide and nucleotide sequences of rat CD4 (W3/25) antigen: evidence for derivation from a structure with four immunoglobulin-related domains. Proc Natl Acad Sci USA 84: 1649–1653
Dalgleish AG, Beverley PC, Clapham PR, Crawford DH, Greaves MF, Weiss RA (1984) The CD4 (T4) antigen is an essential component of the receptor for the AIDS retrovirus. Nature 312: 763–767
Davis SJ, Brady RL, Barclay AN, Harlos K, Dodson GG, Williams AF (1990a) Crystallization of a soluble form of the rat T-cell surface glycoprotein CD4 complexed with Fab from the W3/25 monoclonal antibody. J Mol Biol 213: 7–10
Davis SJ, Ward HA, Puklavec MJ, Willis AC, Williams AF, Barclay AN (1990b) High level expression in Chinese hamster ovary cells of soluble forms of CD4 T lymphocyte glycoprotein including glycosylation variants. J Biol Chem 265: 10410–10418
Doyle C, Strominger JL (1987) Interaction between CD4 and class II MHC molecules mediates cell adhesion. Nature 330: 256–259
Fomsgaard A, Hirsch VM, Johnson PR(1992) Cloning and sequences of primate CD4 molecules: diversity of the cellular receptor for simian immunodeficiency virus/human immunodeficiency virus. Eur J Immunol 22: 2973–2981
Garrett T, Wang J, Yan Y, Liu J, Harrison SC (1993) Refinement and analysis of the structure of the first two domains of human CD4. J Mol Biol 234: 763–778
Harpaz Y, Chothia C (1994) Many of the immunoglobulin superfamily domains in cell adhesion molecules and surface receptors belong to a new structural set which is close to that containing variable domains. J Mol Biol 238: 528–539
Houlgatte R, Scarmato P, Marhomy SE, Martin M, Ostankovitch M, Hatosse S, Vervisch A, Auffray C, Platier Tonnear D (1994) HLA class II antigens and the HIV envelope glycoprotein GP120 bind to the same face of CD4. J Immunol 152: 4475–4488
Hussey RE, Richardson NE, Kowalski M, Brown NR, Chang HC, Siliciano RF, Dorfman T, Walker B, Sodroski J, Reinherz EL (1988) A soluble CD4 protein selectively inhibits HIV replication and syncytium formation. Nature 331: 78–81
Jones EY, Davis SJ, Williams AF, Harlos K, Stuart DI (1992) Crystal structure at 2.8 Å resolution of a soluble form of the cell adhesion molecule CD2. Nature 360: 232–239
Klatzmann D, Champagne E, Chamaret S, Gruest J, Guetard D, Hercend T, Gluckman JC, Montagnier L (1984) T–lymphocyte T4 molecule behaves as the receptor for human retrovirus LAV. Nature 312: 767–768
Kuchel PW, Campbell DG, Barclay AN, Williams AF (1978) Molecular weights of the Thy-1 glycoproteins from rat thymus and brain in the presence and absence of deoxycholate. Biochem J 169: 411–417
Kwong PD, Ryu SE, Hendrickson WA, Axel R, Sweet RM, Falena WG, Hensley P, Sweet RW (1990) Molecular characteristics of recombinant human CD4 as deduced from polymorphic crystals. Proc Natl Acad Sci USA 87: 6423–6427
Lange G, Lewis SJ, Murshudov GN, Dodson GG, Moody PCE, Turkenburg JP, Barclay AN, Brady LR (1994) Crystal structure of an extracellular fragment of the rat CD4 receptor containing domains 3 and 4. Structure 2: 469–481
Langedijk J, Puijk WC, Van HW, Meloen RH (1993) Location of CD4 dimerization site explains critical role of CDR3-like region in HIV-1 infection and T-cell activation and implies a model for complex of coreceptor-MHC. J Biol Chem 268: 16875–16878
Littman DR (1987) The structure of the CD4 and CD8 genes. Annu Rev Immunol 5: 561–584
Maddon PJ, Littman DR, Godfrey M, Maddon DE, Chess L, Axel R (1985) The isolation and nucleotide sequence of a cDNA encoding the T cell surface protein T4: a new member of the immunoglobulin gene family. Cell 42: 93–104
Maddon PJ, Molineaux, SM, Maddon DE, Zimmerman KA, Godfrey M, Alt FW, Chess L, Axel R (1987) Structure and expression of the human and mouse T4 genes. Proc Natl Acad Sci USA 84: 9155–9159
Moebius U, Clayton LK, Abraham S, Harrison SC, Reinherz EL (1992) The human immunodeficiency virus gp120 binding site on CD4: Delineation by quantitative equilibrium and kinetic binding studies of mutants in conjunction with a high-resolution CD4 atomic structure. J Exp Med 176: 507–517

Nag B, Wada HG, Passmore D, Clark BR, Sharma SD, McConnell HM (1993) Purified beta-chain of MHC Class II binds to CD4 molecules on transfected HeLa cells. J Immunol 150: 1358–1364
Pal R, Nair BC, Hoke GM, Sarngadharan MG, Edidin M (1991) Lateral diffusion of CD4 on a human neoplastic T-cell line probed with a fluorescent derivative of the envelope glycoprotein (gp120) of human immunodeficiency virus type 1 (HIV-1). J Cell Physiol 147: 326–332
Reinherz EL, Kung PC, Goldstein G, Schlossman SF (1979) Separation of functional subsets of human T cells by a monoclonal antibody. Proc Natl Acad Sci USA 76: 4061–4065
Rudd CE, Trevillyan JM, Dasgupta JD, Wong LL, Schlossman SF (1988) The CD4 receptor is complexed in detergent lysates to a protein-tyrosine kinase (pp58) from human T lymphocytes. Proc Natl Acad Sci USA 85: 5190–5194
Ryu SE, Kwong PD, Truneh A, Porter TG, Arthos J, Rosenberg M, Dai XP, Xuong NH, Axel R, Sweet RW et al. (1990) Crystal structure of an HIV-binding recombinant fragment of human CD4. Nature 348: 419–426
Ryu SE, Truneh A, Sweet RW, Hendrickson WA (1994) Structures of an HIV and MHC binding fragment from human CD4 as refined in two crystal lattices. Structure 2: 59–74
Schockmel GA, Somoza C, Davis SJ, Williams AF, Healey D (1992) Construction of a binding site for human immunodeficiency virus type 1 gp120 in rat CD4. J Exp Med 175: 301–304
Spellman MW, Leonard CK, Basa LJ, Gelineo I, Van HH (1991) Carbohydrate structures of recombinant soluble human CD4 expressed in Chinese hamster ovary cells. Biochemistry 30: 2395–2406
Triebel F, Jitsukawa S, Baixeras E, Roman RS, Genevee C, Viegals PE, Hercend T (1990) LAG-3, a novel lymphocyte activation gene closely related to CD4. J Exp Med 171: 1393–1405
Tsui P, Sweet RW, Sathe G, Rosenberg M (1992) An efficient phage plaque screen for the random mutational analysis of the interaction of HIV-1 gp120 with human CD4. J Biol Chem 267: 9361–9367
van der Merwe PA, Brown MH, Davis SJ, Barclay AN (1993) Affinity and kinetic analysis of the interaction of the cell-adhesion molecules rat CD2 and CD48. EMBO J 12: 4945–4954
Veillette A, Sleckman BP, Ratnofsky S, Bolen JB, Burakoff SJ (1990) The cytoplasmic domain of CD4 is required for stable association with the lymphocyte-specific tyrosine protein kinase p56lck. Eur J Immunol 20: 1397–1400
Vignali DAA, Doyle C, Kinch MS, Shin J, Strominger JL (1993) Interactions of CD4 with MHC class II molecules, T cell receptors and p56lck. Phil Trans R Soc Lond B 342: 13–24
Wang JH, Yan YW, Garrett TP, Liu JH, Rodgers DW, Garlick RL, Tarr GE, Husain Y, Reinherz EL, Harrison SC (1990) Atomic structure of a fragment of human CD4 containing two immunoglobulin-like domains. Nature 348: 411–418
Webb M, Mason DW, Williams AF (1979) Inhibition of mixed lymphocyte response by monoclonal antibody specific for a rat T lymphocyte subset. Nature 282: 841–843
Weber S, Karjalainen K (1993) Mouse CD4 binds MHC class II with extremely low affinity. Int Immunol 5: 695–698
White RA, Mason DW, Williams AF, Galfre G, Milstein C (1978) T-lymphocyte heterogeneity in the rat-separation of functional subpopulations using a monoclonal antibody. J Exp Med 148: 664–673
Williams AF, Barclay AN (1988) The immunoglobulin superfamily—domains for cell surface recognition. Annu Rev Immunol 6: 381–405
Williams AF, Galfre G, Milstein C (1977) Analysis of cell surfaces by xenogeneic myeloma-hybrid antibodies: differentiation antigens of rat lymphocytes. Cell 12: 663–673
Williams AF, Davis SJ, He Q, Barclay AN (1989) Structural diversity in domains of the immunoglobulin superfamily.Cold Spring Harb Symp Quant Biol LIV: 637–647
Yuen CT, Carr SA, Feizi T (1990) The spectrum of N-linked oligosaccharide structures detected by enzymic microsequencing on a recombinant soluble CD4 glycoprotein from Chinese hamster ovary cells. Eur J Biochem 192: 523–528

# The Structural Basis of CD4 – MHC Class II Interactions: Coreceptor Contributions to T Cell Receptor Antigen Recognition and Oligomerization-Dependent Signal Transduction

R. König[1], S. Fleury[2], and R.N. Germain[2]



## 1 Introduction

During thymic differentiation, precursor T lymphocytes committed to the αβ receptor lineage differentiate along two distinct pathways into mature cells distinguished by the mutually exclusive expression of either the CD4 or CD8 cell surface glycoproteins (Reinherz et al. 1980; Reinherz and Schlossman 1980; Fitch 1986; von Boehmer 1988). Early studies of these two lymphocyte subpopulations

[1]University of Texas Medical Branch, Department of Microbiology and Immunology, Sealy Center for Molecular Science, 301 University Boulevard, Route J-19, Galveston, TX 77555–1019, USA
[2]Lymphocyte Biology Section, Laboratory of Immunology, National Institute of Allergy and Infectious Diseases, National Institutes of Health, 10 Center Drive MSC 1892, Bldg. 10, Rm. 11N311, Bethesda, MD 20892-1892, USA

($CD4^+$ and $CD8^+$) suggested an association between the functional behavior of a T cell and its expression of either CD4 or CD8, with the former associated with B cell helper activity or delayed type hypersensitivity and the latter with cytotoxic activity. As the importance of class I and class II major histocompatibility complex (MHC) proteins in antigen recognition by αβ T cells became apparent, several investigators examined whether the best correlation of CD4 and CD8 expression was with effector function or, alternatively, with specificity of MHC molecule recognition. Expression of CD8 was found to be tightly linked with reactivity to class I MHC molecules, whereas expression of CD4 was associated with reactivity to class II MHC antigens, irrespective of the helper/cytotoxic behavior of the T cells being studied (OKADA and HENNEY 1980; SWAIN 1981; ENGLEMAN et al. 1981; BIDDISON et al. 1982; KRENSKY et al. 1982; MEUER et al. 1982; SPITS et al. 1982; SWAIN 1983; SWAIN et al. 1983; FITCH 1986; BIERER et al. 1989).

It has become clear over the past decade that the relationship between T cell recognition of class I or class II glycoproteins and CD8 vs CD4 expression, respectively, is the result of the affinity of these latter surface molecules for relatively monomorphic portions of the polymorphic MHC proteins. This affinity comes into play both during intrathymic selection of T cell precursors for maturation (positive selection) (VON BOEHMER 1994) and during peripheral activation of mature T cells for development of effector function. CD4 and CD8 are not merely adhesion or accessory molecules, but also contribute to T cell activation by acting as coreceptors that help form a multiprotein signaling assembly with the T cell receptor (TCR) and CD3/ζ complex (JANEWAY et al. 1989; JANEWAY 1992). The coreceptors contribute to intracellular messenger generation by participating in antigen recognition and receptor clustering and through their noncovalent association with the protein tyrosine kinase $p56^{lck}$ (RUDD et al. 1989; VEILLETTE et al. 1988, 1989). This review describes the experimental approaches that have led to the view of CD4 and MHC class II molecules as a receptor-ligand pair and the recent mapping of specific regions of both the CD4 and MHC class II proteins that contribute to their interaction. It concludes with new information suggesting a role for oligomerization of TCR, CD4, and MHC class II in functional signaling of T cells and some speculations on how the various intermolecular affinites among the coreceptor, the T cell antigen receptor complex, and their MHC molecule-associated ligand have evolved to create a highly sensitive and specific recognition and signal transduction apparatus.

## 2 Evidence for MHC Class II Recognition by CD4

### 2.1 Functional Studies

Data showing a correlation between CD4 expression and a T cell's preference for antigen recognition in the context of MHC class II molecules led to the first suggestion that CD4 might interact directly with class II (REINHERZ and SCHLOSSMAN

1980). Additional support for this notion came from studies using monoclonal anti-CD4 antibodies (BIDDISON et al. 1982; REINHERZ et al. 1983), which were found to block a wide variety of antigen-specific T cell responses, such as proliferation, interleukin-2 (IL-2) secretion, cytolysis, and TCR-dependent adhesion, by class II MHC-restricted polyclonal T cells and T cell clones (DIALYNAS et al. 1983; WILDE et al. 1983; SWAIN et al. 1984). An inverse correlation between the apparent affinity of a T cell, based on its antigen sensitivity, and the T cell's susceptibility to functional inhibition by anti-CD4 antibodies (MARRACK et al. 1983) led to the idea that CD4 contributed to T cell activation largely by increasing cell-to-cell avidity, thereby promoting effective T cell receptor occupancy and signaling.

A major problem in these early studies on CD4 function was distinguishing between the required recognition of the class II MHC molecule by the TCR itself and interaction of the CD4 molecule with this same class II molecule. GAY and colleagues used gene transfer and an unusual T cell hybridoma (3DT-52.5.8) that recognizes the murine class I molecule H-2D$^d$ (GREENSTEIN et al. 1984) to dissociate the MHC specificity of the TCR from the function of the CD4 molecule (GAY et al. 1987). In the absence of either CD4 or CD8 expression, the 3DT-52.5.8 hybridoma secreted IL-2 in response to H-2D$^d$-bearing P815 cells. When the 3DT-52.5.8 hybridoma was transfected with a cDNA encoding human CD4, it responded with increased IL-2 secretion upon stimulation with H-2D$^d$-bearing P815 cells that were transfected with cDNAs encoding the human HLA-DR class II MHC molecule. IL-2 secretion by the CD4$^-$ parental line did not differ after stimulation with HLA-DR-expressing or nonexpressing P815 cells. Monoclonal antibodies against CD4 or class II MHC molecules specifically inhibited T cell activation in this system. These findings therefore suggested a direct interaction between CD4 and HLA-DR that augmented T cell activation independent of an interaction between the TCR and MHC class II. Subsequent studies showed that CD4$^+$but not CD4$^-$ 3DT-52.5.8 T cells gave enhanced responses when any of four different class II MHC molecules were incorporated into planar membranes together with H-2D$^d$. Anti-CD4 antibodies inhibited this stimulation (GAY et al. 1988). Because enhanced T cell activation in this experiment was independent of the allele of the class II MHC involved, this result suggested that CD4 recognizes a conserved monomorphic determinant on class II MHC molecules.

## 2.2 Binding Studies

Using a TCR-independent system, DOYLE and STROMINGER (1987) first demonstrated a clear direct interaction between CD4 and class II MHC molecules. These investigators employed an SV40-derived recombinant vector to achieve high levels of human CD4 on the surface of infected CV1 fibroblast cells. After infection, these cells expressed approximately tenfold the number of CD4 molecules at the cell surface as did the human T lymphocyte cell line HPB-ALL. Raji cells, which are of human B lymphoblastoid origin and have high levels of plasma membrane MHC class II molecules, tightly adhered to a monolayer of

CD4-expressing CV1 cells. Other class II-expressing B cells (e.g., DAUDI, JY, and T5-1) also bound to CD4-expressing monolayers, whereas mutant B cells lacking class II MHC expression did not adhere. Monoclonal antibodies against CD4 or MHC class II molecules, but not against class I MHC antigens, inhibited the binding reaction. Quantitative analysis of binding using radiolabeled B cells revealed a direct correlation between MHC class II molecule expression and the amount of binding in the adhesion assay.

Doyle and Strominger concluded that the high level of CD4 expression obtained using the SV40 expression system was crucial for demonstrating stable interactions between CD4 and MHC class II molecules. At physiological levels of CD4, interactions between CD4 and class II proteins alone may be insufficient to mediate cell-cell adhesion, because of the extremely low affinity with which the two molecules appear to bind to each other. To date, it has not been possible to detect measurable interactions between monovalent CD4 and class II MHC molecules in solution. Nonetheless, systems in which one or the other molecule has been immobilized at high density have been successfully employed to determine association/dissociation parameters for the binding between CD4 and class II MHC molecules. Thus, CAMMAROTA and coworkers (1992) determined an equilibrium dissociation constant ($K_d$ value of $3.2 \times 10^{-6}$ *M*) by Scatchard analysis for the interaction between soluble human class II MHC molecules (HLA-DR4) and Sepharose-immobilized human CD4. WEBER and KARJALAINEN (1993) measured the $K_d$ between soluble mouse CD4 coupled to beads and class II MHC molecules on A20 B lymphoma cells. This study gave a $K_d$ of $<10^{-4}$ *M*. These data may indicate that the human CD4 protein has a greater affinity for class II molecules than the mouse homolog, which is consistent with some mutagenesis studies performed subsequently (see below). They also imply that either the binding between CD4 and class II MHC molecules is indeed of extremely, low affinity, or else the conformation of the proteins in their natural environment, namely the cell membrane, differs from that of soluble forms, providing for a stronger interaction. Furthermore, during the binding of class II MHC-bearing cells to $CD4^+$ cells, oligomerization of the interacting molecules may be induced, thereby significantly enhancing the avidity of the interaction.

## 3 General Models for CD4 Function in T Cell Activation

These and numerous other similar studies on CD4-MHC class II interactions gave rise to two alternative models of CD4 function during T cell activation by cell-associated TCR ligands. The *accessory molecule* hypothesis held that TCR-independent physical association between CD4 on the T cell and MHC class II molecules on the antigen-presenting cell (APC) contributed to enhanced cell-cell adhesion. This in turn would increase the opportunity for the TCR to encounter its ligand on the APC surface, thus augmenting occupancy-determined T cell stimulation at a given antigen density, as compared to a responding T cell lacking

CD4 expression. The gene transfer experiments involving the 3DT-52.5.8 T cell hybridoma described in the preceding section are consistent with this model. In this circumstance, the TCR and the CD4 molecule do not recognize/bind to the same molecule. Each recognition event (of class II by the TCR and by CD4) is physically independent, and the CD4 molecule cannot directly enhance specific peptide-MHC molecule interaction with the TCR.

This accessory molecule model, however, cannot explain the strict association of CD4 expression with class II MHC restriction observed in normal T cells. If such a *trans*-active accessory function that contributed to overall intercellular avidity were sufficient for the role of CD4 in T cell signaling, CD4 and CD8 would be functionally interchangeable, because most professional APCs express both class I and class II MHC. Therefore, a T cell that has a class I MHC-specific TCR expressed together with CD4 would obtain the same accessory molecule function as would a T cell with a TCR restricted by class II MHC and paired with CD4. Although experiments such as those with 3DT-52.5.8 and others (Bierer et al. 1989) show that such a *trans*-active function of CD4 or CD8 can be observed, the vast majority of studies report that this effect is substantially weaker than that observed when the MHC class specificity of the TCR and the CD4 or CD8 molecule is the same (see, for example, Miceli et al. 1991). This model also fails to account for the correlation between CD4 or CD8 expression and thymocyte selection based on the MHC class specificity of the TCR.

To explain the strict association of CD4 expression with class II MHC restriction, Janeway and colleagues (1989) suggested an alternative model for CD4 function: the *coreceptor hypothesis*. This model postulates that, aside from any adhesive function, CD4 plays a critical role as a component of the TCR complex required for efficient and proper antigen-dependent signal transduction. The model makes several predictions: (1) that both CD4 and the TCR bind to the same class II MHC molecule; (2) that CD4 physically associates with the TCR on the cell surface upon antigen-stimulation; (3) that the complex of CD4 with the TCR generates a much stronger signal for T cell activation than does the TCR alone; (4) that CD4 can transduce a signal. All these predictions are supported by experimental evidence (Janeway et al. 1989; Janeway 1992; Miceli and Parnes 1993), and it is now generally accepted that the primary physiological function of CD4 relates to its corecognition of the same MHC molecule as the TCR, rather than to its contribution to intercellular avidity.

# 4 Mutational Analysis of CD4 Reveals Multiple Sites of Interaction with Class II MHC Molecules

## 4.1 Studies Using Binding Assays

Site-directed mutagenesis and gene transfer techniques have yielded a substantial understanding of the molecular basis for CD4-MHC class II molecule interactions. The atomic resolution of the three-dimensional structure of crystals consisting of the two NH2-terminal immunoglobulin-like domains (D1 and D2) of CD4 made it possible to relate the results of mutational analyses to the structure of CD4 (Ryu et al. 1990; Wang et al. 1990). The extended, rod-like structure (25×25×125 Å) of all four immunoglobulin-like domains of recombinant soluble human CD4 (Kwong et al. 1990) is consistent with a model in which CD4 spans the TCR-MHC class II complex, and at least the two NH2-terminal domains of CD4 reach to monomorphic regions of class II MHC molecules on the opposing antigen-presenting cell.

Reinherz and colleagues have used a binding assay modeled on the original system described by Doyle and Strominger to investigate the site(s) on CD4 involved in binding to class II molecules (Clayton et al. 1989; Moebius et al. 1992a, 1993). A full-length cDNA encoding human CD4 was subcloned into the eukaryotic expression vector CDM8 (Seed and Aruffo 1987) and used for transient expression studies with COS-1 cells. Human B lymphoblastoid cells expressing high levels of surface class II molecules adhered to monolayers of COS-1 cells transfected with CDM8-CD4, but not to monolayers transfected with CDM8 vector alone (Clayton et al. 1989). Antibodies against CD4 or class II, but not against CD8 or class I, inhibited adherence, demonstrating that cell adhesion was dependent on the binding of CD4 to class II molecules. Because this system does not involve TCR recognition, and because COS-1 cells do not express the CD4-associated protein tyrosine kinase $p56^{lck}$, this model allows analysis of CD4-MHC class II interaction independent of other cellular events potentially initiated by this recognition event (e.g., up-regulation or activation of adhesion molecules, aggregation of CD4-TCR-MHC class II complexes). By the same token, however, the absence of these events limits the utility of this approach, as it may fail to reveal important changes in CD4 interaction with TCR-CD3 proteins upon recognition of MHC class II molecules. Therefore, it cannot take into account possible cooperative effects of TCR and CD4 binding to a single class II molecule (coreceptor function), and it may only reflect the behavior of CD4 lacking cytosolic associations that could have significant effects on the structure or binding properties of the CD4 molecule.

In this assay system, human B lymphoblastoid cells were observed to bind poorly to COS-1 cells expressing mouse rather than human CD4. Based on this observation, 17 individual CD4 mutants were generated that substituted murine for human residues between positions 17 and 164 that differed in a nonconservative manner between the two homologues. This approach presumably minimizes changes in the conformation of CD4 induced by nonconserved replacements, because the substituted amino acids occur in a structurally similar protein. There remains, however, the possibility that the species-specific changes are normally paired with each other, and thus individual mutational changes might still be highly disruptive. In an attempt to avoid this problem, clustered substitution mutants were made in which polymorphic amino acids that are located proximate to each other in the linear CD4 sequence were grouped together in the same mutation (Clayton et al. 1989). Each CD4 mutant was transfected into COS-1 cells and tested for class II MHC binding and for preservation of three antibody epitopes. Three mutations affected the structure of extracellular CD4 domains such that they lost most or all of their reactivity with the monoclonal anti-CD4 antibodies. The other 14 mutants could not be distinguished serologically from wild-type CD4, suggesting that the mutations did not lead to gross conformational changes, although the epitope testing was limited in these studies and some changes might have been missed. Seven of the mutations drastically reduced or eliminated binding of the COS transfectants to class II-expressing B cells. These mutations are located in the CDR1- and CDR2-homologous regions of domain 1, in other loops of domain 1, and in domain 2 of CD4. These data were interpreted to indicate that a rather extended surface area, apparently not confined to a single face of the CD4 molecule, interacted with class II MHC molecules.

More recently, Moebius and coworkers (1993) substituted single amino acid residues in human CD4 with alanine, rather than making species replacement mutations. Some residues that are located in surface grooves, as revealed in the crystal structure of CD4, were replaced by tryptophan to fill the groove and block potential interactions with class II MHC. To minimize structural changes, only exposed side-chains were mutated. The resultant molecules were examined with a larger panel of monoclonal anti-CD4 antibodies, which showed that epitopic loss was confined to the local region of the mutation, consistent with little effect on the overall CD4 structure.Using the cell-cell adhesion assay, residues that affect interaction with class II MHC molecules were identified on all lateral surfaces of domain 1 and on domain 2 close to the interface with the first domain. Alanine substitutions for residues located at the top of the CDR1 and CDR2 loops, in the middle part of domain 2, and in domain 3 did not affect binding to class II MHC. Although the amino acids substituted in this study were not identical to those mutated by Clayton and colleagues, both studies implicated the same regions of CD4 in interaction with class II MHC molecules.

## 4.2 Functional Studies

A similar mutagenesis approach, but with a different assay system, was employed by other investigators to examine this same issue (Lamarre et al. 1989b). Recombinant retroviruses (Peterson and Seed 1988) were used to express wild-type or mutant CD4 on the plasma membrane of the murine $CD4^-$, $CD8^-$ T cell hybridoma 3DT-52.5.8. As mentioned above, this T hybridoma's TCR is specific for the murine class I molecule H-$2D^d$ (Greenstein et al. 1984), and the hybridoma secretes IL-2 in response to H-$2D^d$ (Gay et al. 1987). When exposed to murine fibroblasts transfected with H-$2D^d$ and the class II molecule HLA-DP, T cell hybridomas expressing wild-type CD4 increased IL-2 production by more than 25-fold as compared to stimulation with APC expressing only H-$2D^d$. The increase in IL-2 production was inhibited by antibodies against HLA-DP or CD4, but not against HLA-DR, which was not present on the cells used. In addition, after transduction with CD4, the T cell hybridomas formed rosettes on monolayers of fibroblasts that express both H-2D$^d$ and HLA-DP, and antibodies against CD4 or DP inhibited rosette formation. It is, however, not clear whether the adhesion observed in this system is directly mediated by CD4-MHC class II interactions, or rather results from a CD4-dependent augmentation of TCR signaling that in turn affects the activity of integrins on the T cells and their affinity for surface proteins or attached extracellular matrix molecules on the APC (Dustin and Springer 1989). Together, these several pieces of data demonstrated the specificity of the system for CD4-MHC class II interactions. Although still not capable of revealing the full function of CD4 as a true coreceptor, this system has two advantages over a pure cell-cell adhesion assay: first, CD4-class II MHC interactions can be detected at physiological expression levels of the two molecules, and second, CD4 interaction with components of the T cell's signal transduction apparatus, such as p56$^{lck}$, may participate in the activities being measured.

One set of CD4 variants involved homolog-scanning mutants created by substituting clusters of human CD4 residues with the corresponding amino acids of mouse CD4 (Lamarre et al. 1989a). Using the functional and rosetting assays, Lamarre and colleagues found that 3DT-52.5.8 hybridomas expressing seven out of eight CD4 mutants produced significantly less IL-2 than T hybridomas expressing wild-type CD4 (Lamarre et al. 1989a). Three of these clusters, residing in domains 2 and 3 of CD4, are distinct from those previously identified (Clayton et al. 1989). In four mutants, the rosetting and functional assays yielded consistent results, with both effects of CD4-MHC class II molecule interaction being drastically reduced. Three mutants that were moderately affected in their ability to produce IL-2 upon stimulation with both H-2D$^d$ and HLA-DP (<10 fold reduction) still formed rosettes on double-transfectant monolayers. It appears that not all CD4-MHC class II interactions that promote cell-cell adhesion also induce CD4-dependent enhanced T cell responses (i.e., IL-2 secretion). Thus, it is conceivable that certain types of interactions between CD4 and class II MHC can activate or up-regulate integrin function on the responding T cell without contributing appreciably to increased TCR-dependent signaling for effector functions such as cytokine secretion. A similar dichotomy between cell adhesion and IL-2 production has recently been seen in assays of TCR antagonists, which inhibit intracellular events leading to IL-2 production without preventing surface molecule-dependent increases in cell-cell binding (Ruppert et al. 1993).

In an additional analysis using the same system, Fleury and colleagues (1991) tested a panel of CD4 mutants covering all four extracellular domains of the protein. These mutations consisted of (a) epitope-loss mutations in which one to three residues in domains 1, 2, or 4 were substituted (Peterson and Seed 1988), (b) linker-scanning mutations in which the dipeptide serine-arginine was inserted at predicted hydrophilic sites (Mizukami et al. 1988), and (c) homolog-scanning mutations in which three to eight amino acid residues displaying species polymorphism were substituted with the corresponding murine residues. The structural integrity of mutant CD4 molecules was also assayed by cytofluorometric analysis of anti-CD4 antibody binding using 14 monoclonal antibodies grouped in eight clusters, each defining a different region on the CD4 molecule. This analysis showed that most mutations induced only local structural changes, leaving regions distant to the mutations intact. Assays with this panel of mutant CD4 molecules expressed by 3DT-52.5.8 generated data consistent with interaction of an extended surface on one side of the CD4 molecule with its class II ligand. The CD4 subregions involved in this interaction include the CDR1 and CDR3-like loops of domain 1 and a third region in domain 2 (the FG loop) close to the interface with domain 1. Residues 19, 89, and 165 seemed of particular importance. These three residues are all located on the same face of CD4, and their side chains are exposed to solvent. Additionally, this study implicated a polymorphic region in domain 4 of CD4 in the interaction with class II MHC.

Contrary to other studies using the same CD4 mutant (MOEBIUS et al. 1992a, 1993), FLEURY et al. concluded that the CDR2-like loop did not contribute to interaction with class II MHC molecules, based on the failure of deletion of residues 43–49 in this loop to affect either functional or rosetting assays. A similar deletion mutant (residues 42–49 deleted), however, abrogated CD4-dependent IL-2 secretion and rosetting in a model system in which the TCR and CD4 interact with the same MHC molecule (BOWMAN et al. 1990). Also, substitution of Ile for Phe-43 in the CDR2-like loop abrogated binding in the COS-1/B cell binding assay (MOEBIUS et al. 1993). Both deletion mutations and single-site substitutions in the CDR2-like region of CD4 affect the conformation of the molecule as judged from the loss of several antibody epitopes (BOWMAN et al. 1990; FLEURY et al. 1991; MOEBIUS et al. 1993), and, remarkably, the pattern of antibody binding to the mutant CD4 varies with the cell type in which the protein is expressed (FLEURY et al. 1991). If these staining results reflect alternative conformations of CD4 in the different test systems, such structural variations may result in a distinct contribution of the CDR2 region to CD4-class II molecule interaction in each case. Alternatively, this region may participate in oligomerization events required for strong direct adhesion or coreceptor function, but not for accessory molecule function in the 3DT-52.5.8 assay (see below). Clearly, further work is needed to resolve these discrepancies, and to help formulate a consistent model of CD4 structure-function relationships.

## 4.3 Summary of CD4 Structure-Function Studies

Taken together, these mutational analyses have identified multiple regions of CD4 that affect either direct binding to class II MHC molecules or downstream functions dependent on interaction with class II molecules. An extended region of functional significance is formed on one side of domain 1 by residues in the A-strand, the A/B-loop, the B-strand, the beginning of the CDR1-like B/C-loop, the CDR3-like F/G-loop, and the D-strand, as well as residues in the D/E-loop (Asp-63) and at the end of the E-strand (Lys-72) (CLAYTON et al. 1989; MOEBIUS et al. 1993). This active region may extend into domain 2, where the F/G-loop, a part of the G-strand, and the C/C'-loop are also implicated in class II-dependent CD4 function. On the other side of domain 1, the C-strand, the C/C' and C''/D-loops, and possibly the CDR2-like C'/C"-loop and C"-strand form another region affecting class II-dependent activity. On the same side of CD4, residues in the A-strand, and the A/B and B/C-loops of domain 2 also contribute to function. These data imply the existence of two extended regions located on two spatially discrete faces of the CD4 molecule that are involved in MHC class II-dependent CD4 function (Fig. 1A,B). Models that account for these unexpected data will be considered below after describing emerging information on the sites on class II molecules involved in interaction with CD4.

In addition to these two extended areas, Sekaly's group has reported disturbance of CD4 function in 3DT-52.5.8 cells following mutation of domain 3 (LAMARRE et al. 1989a) and domain 4 (FLEURY et al. 1991) by substitution of clusters of species-polymorphic mouse CD4 residues for the corresponding residues in human CD4. Using single-site alanine substitutions, MOEBIUS and colleagues (1993) also introduced mutations into domain 3 of CD4, although in different residues from those altered in the experiments of LAMARRE. They, however, failed to detect an effect on CD4 function using the cell-cell adhesion assay.

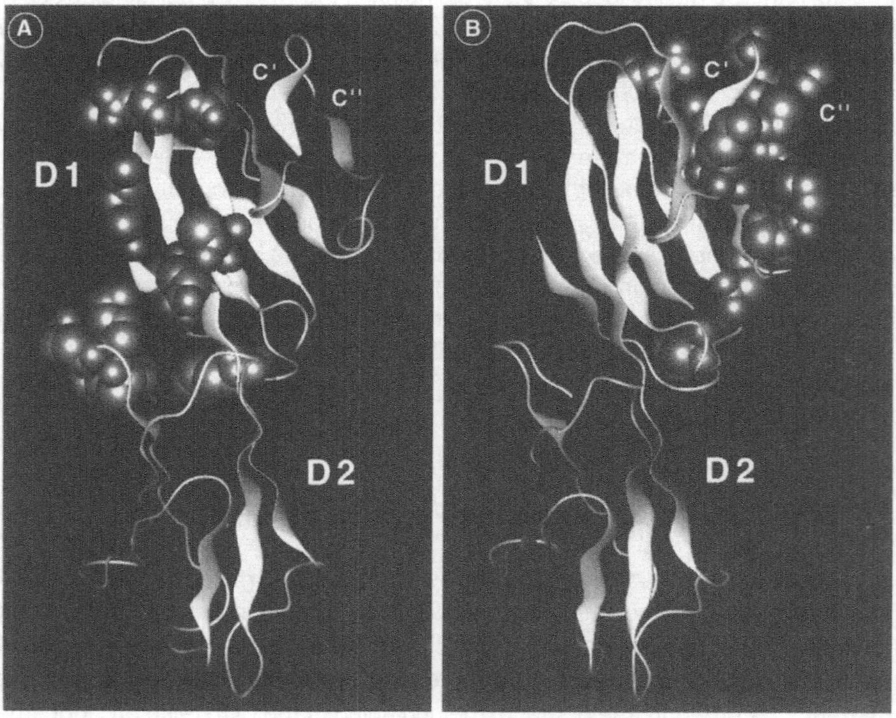


**Fig. 1 A,B.** **A** Side view of the D1 and D2 domains of human CD4 (ribbon drawing), with the residues involved in functional interaction with MHC class II molecules but not gp120 binding shown in *dark gray* atomic outline. **B** Side view of the D1 and D2 domains of human CD4 (ribbon drawing), with the residues involved in functional interaction with MHC class II molecules and overlapping with the region affecting gp120 binding shown in *dark gray* atomic outline. *D1*, $NH_2$-terminal domain of CD4; *D2*, second domain of CD4; *C' and C"*, strands of the D1 region

The dimensions of the molecules involved in CD4-MHC class II interaction raise questions about the likelihood that regions in the membrane-proximal domains 3 and 4 of CD4 participate directly in the interaction with class II MHC. The CD4 molecule is about 125 Å long (KWONG et al. 1990), whereas the height of a class II MHC molecule is about 73 Å (BROWN et al. 1993) and that of the TCR may be 60–70 Å (considering that both the α and the β chains consist of two immunoglobulin-like domains). Therefore, the D3 and D4 domains of CD4 would be unlikely to reach beyond the TCR to contact class II MHC molecules if perpendicular orientations are considered. If this proves to be the case, how then can mutations in D3 and D4 affect CD4 function? One possibility is that the mutations may induce structural changes in distant regions of CD4, indirectly disturbing interactions with class II MHC and/or TCR/CD3 components. In this regard, it is noteworthy that the mutation in D4 that affected CD4 function in the 3DT-52.5.8 assay also affected eight out of 14 antibody epitopes in six out of eight clusters (FLEURY et al. 1991). Alternatively, the signaling function of CD4 may depend on protein-protein interactions, either between CD4 molecules or between CD4 and TCR/CD3 components (VIGNALI et al. 1993), and this may involve the membrane-proximal D3 and D4 domains. Although there is no direct physical evidence for CD4 dimerization, this consideration is consistent with observations suggesting that CD4 and TCR molecules associate in activated T cells in a 2 : 1 ratio (ANDERSON et al. 1988; ROJO et al 1989; SAIZAWA et al. 1987).

## 5 Class II MHC Molecules and the HIV Envelope Glycoprotein gp120 May Bind to the Same CD4 Site

The HIV envelope glycoprotein gp120, which binds to CD4, can inhibit adherence of class II-bearing B cells to human CD4 (CLAYTON et al. 1989; ROSENSTEIN et al. 1990). Several laboratories have analyzed the capacity of mutant CD4-expressing transfectant cells to bind gp120. Some assay systems used co-immunoprecipitation of CD4 and gp120 to detect association (CLAYTON et al. 1988; MIZUKAMI et al. 1988), or flow cytometry to measure equilibrium binding (CHANDAU et al. 1988; BOWMAN et al. 1990; BRODSKY et al. 1990; CLAYTON et al. 1989; MOEBIUS et al. 1992b) or the kinetics of CD4-gp120 interactions (MOEBIUS et al. 1992b). Other systems employed saturation binding of radiolabeled gp120 to transfectant cells expressing CD4 (ASHKENAZI et al. 1990; FLEURY et al. 1991; LAMARRE et al. 1989a). The principal region of gp120 binding is located on and around the CDR2-like C'C" ridge on domain 1 of CD4 and encompasses a contiguous surface-exposed area of about 900 Å$^2$. Of particular importance for the binding of gp120 are four positively charged amino acid residues (three lysines and one arginine) that partially surround Phe-43, which is located in the center of this area (MOEBIUS et al. 1992b). Additional evidence implicating these positively charged residues in the interaction with gp120 comes from the observation that three negatively charged residues in gp120 are critical for binding to CD4 (Olshevsky et al. 1990). These residues are highly conserved amongst all HIV-1 isolates.

As discussed above, the participation of the CDR2-like loop of CD4 in interaction with class II MHC is still controversial. If this region represents a partial overlap of the binding sites on CD4 for gp120 and class II MHC molecules, direct competition of gp120 with class II MHC molecules for CD4 binding could explain the block exerted by gp120 on the association between class II-bearing and CD4-expressing cells (Clayton et al. 1989; Rosenstein et al. 1990). Alternatively, steric hindrance by gp120 of class II-binding to CD4 at a separate site would achieve the same effect, as would induced structural changes in CD4. The equilibrium dissociation constant for gp120-binding to CD4 ($\sim 10^{-9} M$) is several orders of magnitude stronger than that for class II-binding to CD4. Inhibition of CD4-class II interactions by gp120 may therefore in part account for HIV-induced immunodeficiencies by blocking development and function of $CD4^+$ T cells.

# 6 Mapping of the Sites on MHC Class II Molecules Involved in Interaction with CD4

## 6.1 Studies of the Polymorphic α1 and β1 Domains

A variety of mutation and transfection strategies have been used to analyze the reciprocal structure-function relationship, namely, the regions of MHC class II molecules involved in interaction with CD4. In an effort to examine the ability of T cells to recognize a single class II MHC domain, McCluskey and colleagues constructed recombinant murine class II/class I MHC genes (McCluskey et al. 1985, 1988). These recombinant genes allowed the isolated expression of the β1 or α1 domains of $A^k$ anchored by the class I α3, transmembrane, and cytoplasmic regions on the surface of mouse L cells. Both class II regions could elicit specific allogeneic responses by cytotoxic $CD4^+$ T cell lines (Golding et al. 1985; McCluskey et al. 1988). Furthermore, Golding and coworkers (1985) demonstrated an inhibitory effect of anti-CD4 antibodies on the cytolytic activity of an anti-$A^k$β1 allospecific T cell line. This result was interpreted to show that either CD4 function in T cell interactions can be mediated solely by the β1 domain and does not require the presence on target cells of any other class II MHC domains, or that the allospecific T cell line in question does not require CD4-MHC class II interactions for responding to the alloantigen, and that the effect of anti-CD4 antibodies results from the induction of a CD4-mediated negative signal. Indeed, such negative signaling mediated by CD4 has been demonstrated (Julius et al. 1993). Similar experiments using anti-CD4 antibodies to block the cytotoxic activity of anti-$A^k$ α1 allospecific T cells were not performed.

More recently, peptide competition experiments and mutagenesis of the β1 domain have provided more direct evidence that this region of MHC class II molecules may directly participate in interaction with CD4, as suggested by GOLDING et al. Synthetic peptides corresponding to amino acid residues 35–46 of HLA-DQ1, DR1, or DR2 inhibited conjugate formation between $CD4^+$ T lymphocytes and EBV-transformed B cells (MAZEROLLES et al. 1990). Another group used a similar peptide to compete for binding of radiolabeled soluble HLA-DR to CD4-transfected HeLa cells (NAG et al. 1993). Mutational analysis of the β1 domain of HLA-DR1 suggests a contribution of the solvent-exposed loop formed by amino acid residues 41–55 to interaction with CD4 as assayed by cell binding (Brogdon and Doyle, personal communication). The CD4-binding region on the class II β1 domain might interact with amino acid residues in the FG-loop of the second domain of CD4, thereby stabilizing the class II MHC-CD4 interaction.

## 6.2 Studies Indicating a Critical Role of the Immunoglobulin-Like β2 Domain

The further characterization of MHC class II regions controlling functional interactions with CD4 was greatly aided by the observation that murine CD4 only poorly interacts with human class II MHC molecules in some assays (CLAYTON et al. 1989; LAMARRE et al. 1989a), whereas human CD4 interacts well with class II MHC molecules from either species (VON HOEGEN et al. 1989). This species-biased interaction between CD4 and class II MHC molecules prompted several groups of researchers to attempt to identify domains on class II MHC that might mediate CD4 binding. In one study, purified $CD4^+$ BALB/c splenocytes were stimulated with the BALB/c-derived B cell line A20.J (haplotype $H\text{-}2^d$). The $CD4^+$ T cells responded with IL-2 secretion and clustering only when the A20.J cells expressed an allogeneic or xenogeneic MHC class II molecule following transfection (VIGNALI et al. 1992). These investigators observed a strong allogeneic response to $A\alpha^b$-$A\beta^b$-transfected A20.J cells, but a much weaker xenogeneic response to human DR3-transfected A20.J cells. Substituting β2 of DR3 for the homologous domain of $A\beta^b$ reduced T cell responses to the same level achieved by stimulation with DR3. This result confirmed previous reports of a species bias in the interaction between CD4 and class II MHC and implicated the β2-domain in interaction with CD4. Substitution of the β2-domain of $A^b$ for the homologous domain in DR3, however, failed to restore T cell responses. This result suggested that regions in the class II MHC α-chain or the β1-domain may also participate in the interaction with CD4 in a species-specific manner, as suggested by the data of Golding et al., although the absence of only the species-matched β2 region can reduce CD4 function to the same level observed upon replacement of all regions.

A second experimental system employed an $A\alpha^kA\beta^b$-restricted T Cell hybridoma specific for peptide 46–61 of hen egg lysozyme (HEL). A CD4-loss variant was selected and stimulated with A20.J cells transfected with either $A\alpha^bA\beta^b$ or a chimeric $A\alpha^bA\beta^b$ molecule carrying the β2-domain of DR3. The CD4-negative T hybridoma responded weakly to both APCs, but did not discriminate between wild-type or chimeric class II MHC. This showed that peptide binding and class II MHC/peptide interaction with the TCR was not affected by the introduction of a foreign β2-domain. In contrast, when the T cell hybridoma was transfected with mouse CD4, it responded much more vigorously to stimulation with either class II molecule than the CD4-negative cell line, but also strongly discriminated between wild-type $A\alpha^bA\beta^b$ and the chimera. This result again implicated the β2-domain in interaction with CD4.

In another set of studies, human $CD4^+$ T cell clones (LOMBARDI et al. 1991) or purified human $CD4^+$ peripheral blood lymphocytes (PBLs) (RAMESH et al. 1992) were stimulated with fibroblasts expressing human HLA-DR1 class II molecules or chimeras between DR1β and mouse Eβ. To generate the chimeric molecules, the NH2 -or COOH-terminal half of the β1 domain, the β2 domain, or combinations of two of these regions from Eβ were substituted for the corresponding regions in DR1, and the resulting recombinant β-chains were paired with either a wild-type DR1α or Eα chain. Neither the T cell clones nor the $CD4^+$ PBLs responded to the xenoantigen EαEβ. Both, however, proliferated vigorously after stimulation with the alloantigen DR1. T cells that were dependent on CD4 function for a response (i.e., that could be blocked by anti-CD4 antibody) also required the presence of a DR1 β2-domain. Substituting either or both halves of the Eβ1 domain for the corresponding DR1 region eliminated allorecognition by all but one T cell clone, whether or not allorecognition could be blocked with anti-CD4 antibody. This result suggests that polymorphic residues in the β1 domain are recognized by the TCR or that mismatches between the α chain and the β1 domain induce conformational changes that prevent allorecognition independent of an effect on the interaction with CD4. As was the case in the experiments of Vignali et al., the presence of a human β2 domain alone could not restore full allorecognition by $CD4^+$ PBLs.

More precise localization of regions of class II critical for CD4 function came from mutagenesis and transfection studies using a coreceptor-type assay (KÖNIG et al. 1992). One strategy was based on the previously discussed species bias in the interaction between CD4 and class II MHC (CLAYTON et al. 1989; LAMARRE et al. 1989a; VON HOEGEN et al. 1989). Amino acid residues that were conserved amongst all alleles within one species but displayed interspecies diversity were replaced individually in mouse class II α and β chains with the corresponding residues from human class II. A second strategy was based on the structural and functional homologies between class I and class II MHC molecules (BJORKMAN et al. 1987; BRAUNSTEIN and GERMAIN 1987; BROWN et al. 1993; GERMAIN 1994), and the reported binding site for CD8 on class I MHC molecules in the 223–229 loop of the α3 domain (POTTER et al. 1989; SALTER et al. 1990). Because MHC class I and class II molecules perform similar functions, it seemed likely that CD4 might interact with class II MHC through a site located in a region structurally analogous to the CD8-binding site on class I MHC. The crystal structure of class II MHC molecules had not yet been resolved, so a computer model of the murine class II MHC molecule $A\alpha^dA\beta^d$ based on the crystallographic coordinates of the human class I protein HLA-A2 was generated (KÖNIG et al. 1992). Both the class II α2 and β2 domains were modeled on the backbone of the class I α3 domain, because sequence alignments favored this representation over the possible alternative of modeling the class II α2 domain on class I-associated β2-microglobulin. In both the class II α2 and β2 domain models, an exposed segment analogous in position to the CD8-binding loop in the class I α3 domain was identified, involving amino acid residues α2 (125–133) and β2 (137–143), respectively. Alanine-scanning mutagenesis was then employed to evaluate the contribution of individual amino acids in both segments to interaction with CD4.

These two sets of mutant class II molecules were expressed in fibroblasts by gene transfection in the form of heterodimers with one wild-type chain and one mutant chain. Proper expression and serologic behavior indicated that the mutations did not have any gross effect on class II structure. Direct peptide binding experiments also showed that these mutant class II molecules bound antigen in a quantitatively normal fashion. The functional assay for CD4 interaction with MHC class II molecules utilized several T cell hybridomas, some of which constituted sets of T cells all having the same TCR but expressing either human CD4, mouse CD4, mouse CD8, or no coreceptor at all. Such a panel of T cells allows testing the stimulatory potential of wild-type versus mutant class II molecules for cells with and without CD4. By comparison of the relative responses obtained with these combinations of class II and coreceptor expression, one can determine whether a class II mutation selectively affects interaction with CD4 or alters antigen presentation/TCR recognition. Thus, a class II mutation eliminating CD4 function but not affecting T cell recognition should stimulate $CD4^-$ negative T cells to the same extent as wild-type class II. Such a mutant class II molecule should stimulate $CD4^+$ T cells to the same extent as $CD4^-$ T cells, but should elicit a lower response than wild-type class II from these same $CD4^+$ T cells.

Using this assay, several amino acid residues in the β2 137–143 segment were identified in which either species-biased substitutions or, more dramatically, alanine replacements interfered with CD4 coreceptor function (Fig. 2). Changes in class II β chain residues Glu-137, Val-140, Gly-141, and Val-142 showed the strongest impact on CD4 function, with alanine replacement of the absolutely conserved Glu-137 completely eliminating mouse CD4 function and drastically reducing activity with human CD4 (König et al. 1992). By substituting species-conserved amino acid residues in mouse class II with the corresponding human residues, we could identify two amino acids that may account partly for the species bias in the interaction between class II MHC and CD4. One mutation involved substitution of Ser-143 in the putative CD4-binding loop to valine. Although this alteration did not diminish function with mouse CD4, function with human CD4 was significantly improved. Another mutation changed Asn-110 to glutamine, adding a methylene group to increase the size of the side chain. This mutation drastically, but not completely, decreased function of the mutant $A\alpha^dA\beta^d$ protein with mouse CD4, whereas it did not affect function with human CD4. Residue 110 is located on the tip of a surface-exposed loop on the same face of the β2 domain as is segment 137–143 (Brown et al. 1993).

Independently, Cammarota and colleagues (1992) identified a CD4-binding site within the human DRβ2 sequence (residues 134–148) using a binding assay

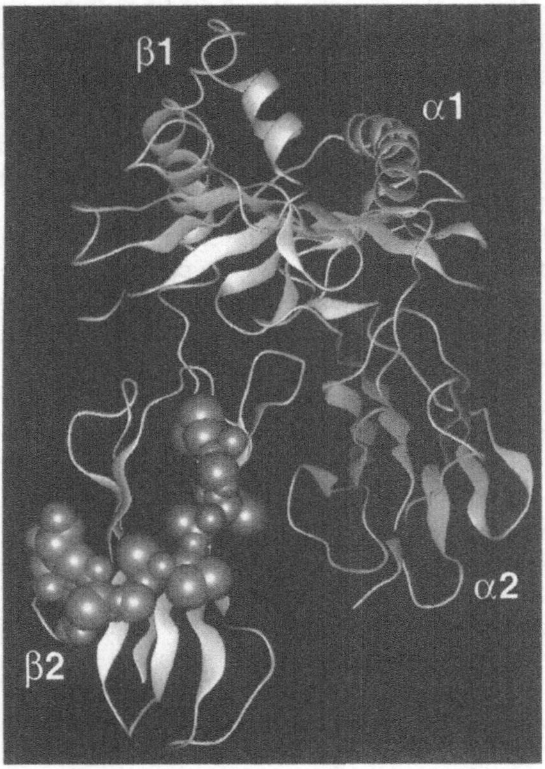


**Fig. 2.** Side view of the human HLA-DR1 class II molecule (ribbon drawing), with the 137–147 loop in the β2 domain that controls interaction with CD4 shown in *dark gray* atomic outline. *α1, α2, β1, β2,* domains of the MHC class II molecule

between CD4 and HLA-DR4-derived peptides. To establish this assay, an affinity matrix was created by immobilizing recombinant soluble CD4 on Sepharose 4B. Solubilized DR4 αβ heterodimers bound specifically to this matrix, as did the isolated DR4 β chain. A series of overlapping peptides spanning the entire length of the DR4 β2 domain was then synthesized and individual peptides were repeatedly passed over a CD4 affinity column. The column was then washed, and bound peptide eluted at low pH. Only two of 11 15-mer peptides bound to immobilized CD4, peptide 134–148 and, with considerably less potency, peptide 138–152. This result demonstrated the importance of NH2-terminal residues in peptides 134–148 for binding to CD4 and is consistent with the data obtained by mutational analysis. A different peptide analog of this same region was more recently shown to inhibit binding of solubilized class II to CD4-bearing cells (Nag et al. 1993).

Evidence for (an) additional site(s) in the β2 domain that may modulate class II MHC molecule interaction with CD4 has been obtained by Fleury and colleagues using the 3DT-52.5.8 functional assay described above (Fleury et al., in press). This study was based on the observation that, while four different isotypic genes may encode HLA-DR β chains (TROWSDALE et al. 1991), the majority of class II-restricted T cell responses are restricted by products of only one of these genes (OTTENHOFF et al. 1985). To determine the contribution of allelic and isotypic polymorphisms to the interaction between CD4 and HLA-DR molecules, Fleury and coworkers stimulated a human CD4 transfectant of the 3DT-52.5.8 hybridoma with mouse L cell transfectants that expressed comparable levels of various DR alleles and isotypes together with the TCR ligand H-2D$^d$. These experiments revealed a hierarchy in the capacity of different DR molecules to enhance T cell responses, indicating differential abilities to interact with CD4. Because all DR molecules share the same α chain, the observed differences must be determined by polymorphisms in the β chain outside of the conserved 137–143 region.

# 7 Coreceptor Function and the Role of Oligomerization of Class II MHC-CD4-T Cell Receptor Complexes in Antigen-Specific T Cell Activation

## 7.1 Coreceptors Help Preserve T Cell Antigen Recognition Specificity While Augmenting Antigen Sensitivity

Both the functional data and the structural studies described above are consistent with the coreceptor model for CD4 function. CD4 thus augments antigen-specific stimulation of T cells best when both the CD4 molecule and the TCR are able to interact with the same peptide-MHC molecule ligand, and the major binding site for CD4 on MHC class II molecules identified by mutagenesis and peptide inhibition studies lies in a region of the class II molecule that allows simultaneous TCR recognition of the peptide-occupied polymorphic α1 and β1 domains. An intriguing question is why antigen-specific activation of αβ TCR-bearing T cells has evolved to depend on this cooperative recognition of ligand by physically independent molecules (CD4 and TCR).

In contrast to immunoglobulins that recognize a wide diversity of ligand structures, αβ T cells participate in adaptive immune responses by recognizing short peptides bound to a common framework consisting of the surrounding MHC class I or class II molecule (GERMAIN 1994). A TCR must discriminate among thousands of sometimes subtly different peptides bound to identical MHC molecules expressed on a single cell surface. The highly related nature of these ligands for the αβ TCR places certain constraints on the recognition mechanism. High affinity receptors discriminate well between diverse structures, but poorly between related structures, due to the high residual affinity of such receptors for ligands similar to those for which they have optimal fit. Because of this feature of receptor-ligand interactions, αβ TCR cannot be of very high affinity, as this would reduce their specificity of recognition and their ability to distinguish between MHC molecules occupied with peptides of similar structure. Low or moderate affinity receptors have a higher intrinsic discriminatory capacity for structurally related ligands, because as receptor affinity for a variant of the "true" ligand declines, the possibility of occupancy adequate for signal generation rapidly diminishes at modest ligand concentrations. Consistent with these notions, emerging data on the affinity of αβ TCR for peptide-MHC molecule complexes reveal $K_d$s in the $10^{-5}$–$10^{-7}$ range (Corr et al. 1994; Matsui et al. 1991; Weber et al. 1992).

Although modest TCR affinity aids in discriminatory recognition of similar peptide-MHC molecule complexes, this characteristic also limits the sensitivity of the response system. If only a modest affinity TCR and its ligand were involved, a high concentration of either TCR, ligand, or both would be needed to achieve adequate occupancy for effective signaling. Yet the T cell activation system must be very sensitive, to ensure effector responses at the low concentrations of antigen available early after infection. The coreceptor system is the solution to the problem of maintaining high sensitivity and high specificity in the T cell recognition of peptide-MHC complexes. In effect, the CD4 coreceptor is an integral part of the TCR recognition unit, which when removed physically from proximity to the αβ combining site limits the affinity of the receptor for peptide-MHC molecule complexes. This allows high specificity as the TCR scans through the many potential MHC molecule ligands on a cell surface. At the same time, the affinity of the coreceptor alone for MHC molecules is very low, preventing appreciable interactions in the absence of TCR recognition. Once such specific recognition takes place, however, the ability of the coreceptor to simultaneously engage the recognized peptide-MHC molecule and the attached TCR complex results in both a stabilization of coreceptor-MHC molecule binding and a major increase in the effective affinity of the TCR. By occuring after the initial interaction of an αβ TCR with a suitable peptide-MHC complex, this affinity increase can take place without a loss of specificity. The added stability of interaction thus confers high sensitivity on the response system and enables low ligand densities to achieve stimulatory occupancy levels. This sequential, cooperative mechanism permits low numbers of specific ligands on a cell surface (<100–500 out of $>10^5$) MHC molecules (Demotz et al. 1990; Harding and Unanue 1990) to activate the TCR signal transduction machinery to stimulatory levels. A further safeguard is built into the system by associating the $p56^{lck}$ kinase with the CD4 coreceptor, such that signal generation is also dependent on cooperative interactions and not fully inherent in the clonotypic recognition apparatus (the αβ TCR) itself.

## 7.2 Evidence for CD4-Related Oligomerization Events

Although this scheme for high specificity and sensitivity of antigen recognition by αβ T cells accounts in a simple way for the existence of coreceptor molecules, it does not explain how signal transduction upon antigen recognition actually occurs. Virtually all transmembrane receptor systems that rely on tyrosine kinase function to initiate intracellular events (as does the TCR) depend on homo- or heterodimerization of receptor subunits to initiate second messenger generation (ULLRICH and SCHLESSINGER 1990). This minimal model for receptor signaling does not seem to fully describe the properties of the αβ TCR complex, as a large literature shows that simple bivalent cross-linking of the TCR is insufficient for induction of downstream activation events (ASHWELL and KLAUSNER 1990). Rather, data from antibody studies (RATCLIFFE et al. 1992) and investigation using multivalent antigen ligands point to a need for higher order oligomerization for such signaling to occur (SYMER et al. 1992). Yet available data indicate that the TCR/CD3 complex is monovalent (SAITO et al. 1987). How then is univalent peptide-MHC molecule recognition translated into the oligomerization necessary for cell activation?

One possible answer is that the association of the CD4/p56$^{lck}$ kinase complex with the TCR/CD3/ζ complex constitutes the relevant oligomerization event. This possibility, however, is made less likely by the data just cited concerning the higher order structures apparently involved in TCR signal transduction and by the fact that CD4 coreceptor function is preserved when the kinase activity of associated p56$^{lck}$ is eliminated by site-directed mutagenesis (XU and LITTMAN 1993).

An interesting alternative comes from very recent experiments based on the original modeling studies suggesting that both the class II α2 and β2 domains might each contain a region analogous to the CD8 binding loop in class I MHC. The mutational analysis and functional testing of mutants in these regions not only revealed the involvement of the β2 domain 137–143 loop in class II-dependent CD4 function, but also has provided evidence for a lesser, but still significant, contribution of the α2 125–133 segment to this activity (König et al. in press). These two regions of the class II molecule lie on opposite faces on a single class II heterodimer, and it does not appear possible for a single CD4 molecule to simultaneously contact both sites (Fig. 3). Therefore, the ability of mutations in either the β2 site or the α2 site to affect CD4 function suggest that both participate in coreceptor activity and that this must involve some form of CD4-dependent oligomerization which allows contact with these sites on at least two distinct MHC class II heterodimers. It should be noted that these new data suggesting a critical role for oligomerization in TCR-mediated signaling are consistent with the long-standing model which postulates minimal signal transduction units called "immunons" that involve the association of multiple (8–12) individual receptor units (DINTZIS et al. 1976).

Several different organizations of the TCR, MHC class II, and CD4 can be imagined that accommodate the mutagenesis data on CD4 sites affecting functional interaction with class II molecules and the data on multiple class II sites controlling interaction with CD4. The simplest model involves CD4 acting as an adapter between individual MHC class II heterodimers (Fig. 4A). One face of CD4 would bind to the β2 site and one would bind to the α2 site of another heterodimer. A second CD4 molecule could then bind to the β2 site of this

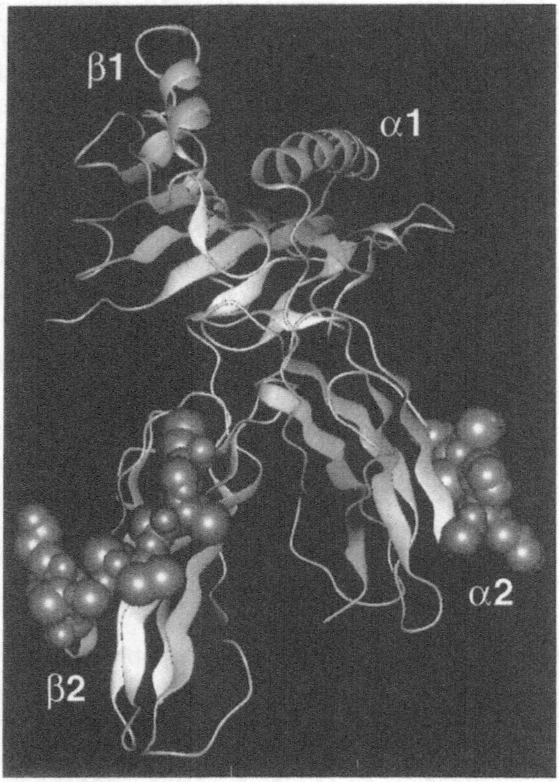


**Fig. 3.** Side view of the human HLA-DR1 class II molecule (ribbon drawing), with both the 137–147 loop in the β2 domain and the 125–131 region of the α2 domain that affects interaction with CD4 shown in *dark gray* atomic outline. α1, α2, β1, β2, domains of the MHC class II molecule

second class II molecule and so on, forming a chain of CD4 and MHC class II molecules. Such binding would only be stable when the CD4 molecule also was bound to a TCR interacting with the same MHC molecule, thus stringing together the TCR/CD3/ζ complexes into oligomers suitable for signal transduction. In this view, the disruption of CD4-MHC class II functional interaction by mutations involving multiple faces of the CD4 molecule would reflect disruption of two distinct but functionally interdependent binding sites on a single CD4 protein. This hypothesis could explain the discrepancy between the data using transfected 3DT-52.5.8 cells and either pure binding or coreceptor assays with mutations in the gp120 binding region of CD4. Because the 3DT-52.5.8 model does not involve MHC class II-dependent oligomerization like a true coreceptor system, nor depend on optimal avidity of interaction as does the COS cell binding assay, one of the two CD4 binding sites for class II molecules may be dispensable under these conditions.

A second class of models arises from the three-dimensional structure of class II MHC crystals (Brown et al. 1993). The structures of class I and class II

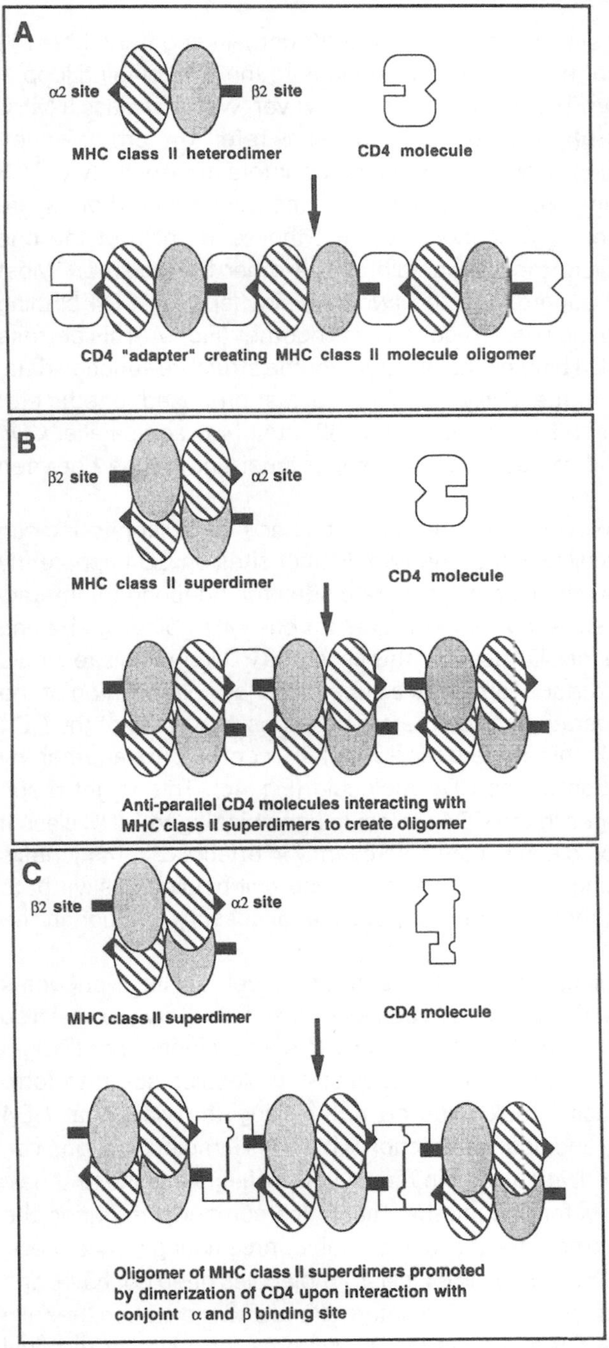


**Fig. 4A-C. A** Adapter model of CD4 function. CD4 links individual class II MHC heterodimers to one another by bridging between the α and β chain sites on different class II molecules. **B** MHC class II superdimer oligomerization by paired CD4 molecules each with separate sites for the MHC class II α and β target regions. **C** MHC class II superdimer oligomerization by CD4 molecules with a single binding site for a conjoint α2/β2 target region and a separate dimerization interface

MHC molecules are very similar, and the critical β2 domain segment 137–143 does indeed lie in a solvent exposed loop analogous to the CD8 binding loop in the MHC class I α3 domain. The major surprise, however, was that class II MHC crystallized as a dimer of αβ heterodimers – we shall refer to them as superdimers– with an orientation allowing simultaneous interaction with two TCR molecules. The membrane-proximal domains of the two heterodimers are oriented in the superdimer in a manner such that the α2 domain of the one heterodimer closely contacts the β2 domain of the other heterodimer. Most interestingly, in a class II superdimer, the two spatially separate CD4 binding sites in the α2 and β2 domains are brought close together (Fig. 5). This permits either of two models for CD4 binding that account for the structure-function data. One possibility is that, as in the "daisy chain" model just proposed, one face of CD4 binds to the α2 site and the other to the β2 site. Two anti-parallel CD4 molecules might then bind to each side of a superdimer and bridge between superdimers (Fig. 4B).

It is questionable, however, whether this postulated molecular packing can occur. An alternative possibility is that the two distinct sites on CD4 apparently involved in MHC class II binding are not separate sites for independent interaction with the two class II sites, but rather represent one MHC binding site and one dimerization interface for CD4 itself. In this case, the CD4 binding site would interact with the conjoint surface formed in a superdimer by juxtaposition of the α2 and β2 sites of two separate heterodimers. The opposite surface of the CD4 molecule would then link this superdimer-CD4-TCR complex to another by promoting homodimerization of the CD4 molecule (Fig. 4c). This might result from conformational changes in the CD4 molecule upon binding to MHC class II molecules. This model would explain the discrepancy in effects of mutagenesis of the CDR2 site by arguing that the coreceptor and cell binding assays both depend critically on this dimerization effect, but that accessory function in the 3DT-52.5.8 assay does not.

One especially appealing feature of this latter model is that it suggests parallels between the CD4 and CD8 coreceptor systems that otherwise seem so different. Much has been made of the fact that CD8 is a covalent dimer (Parnes 1989), whereas CD4 is a monomer, and that class II molecules seem to form dimers in crystals but class I molecules do not. Taking the view that CD4 dimerizes, but only weakly and upon interaction with class II molecules, one can argue that the class II dimerization seen in the crystals reflects a higher intrinsic affinity of class II molecules for one another that helps compensate during the formation of signaling oligomers for the lack of stable, preexisting CD4 dimers. Conversely, because CD8 is a stable dimer, class I molecules need not have, and do not show, evidence of such contributory intermolecular dimerization. Seen in this way, both the class I-CD8 and class II-CD4 systems form similar terminal signaling complexes, but depend on a different distribution of intermolecular binding affinities to accomplish this goal.

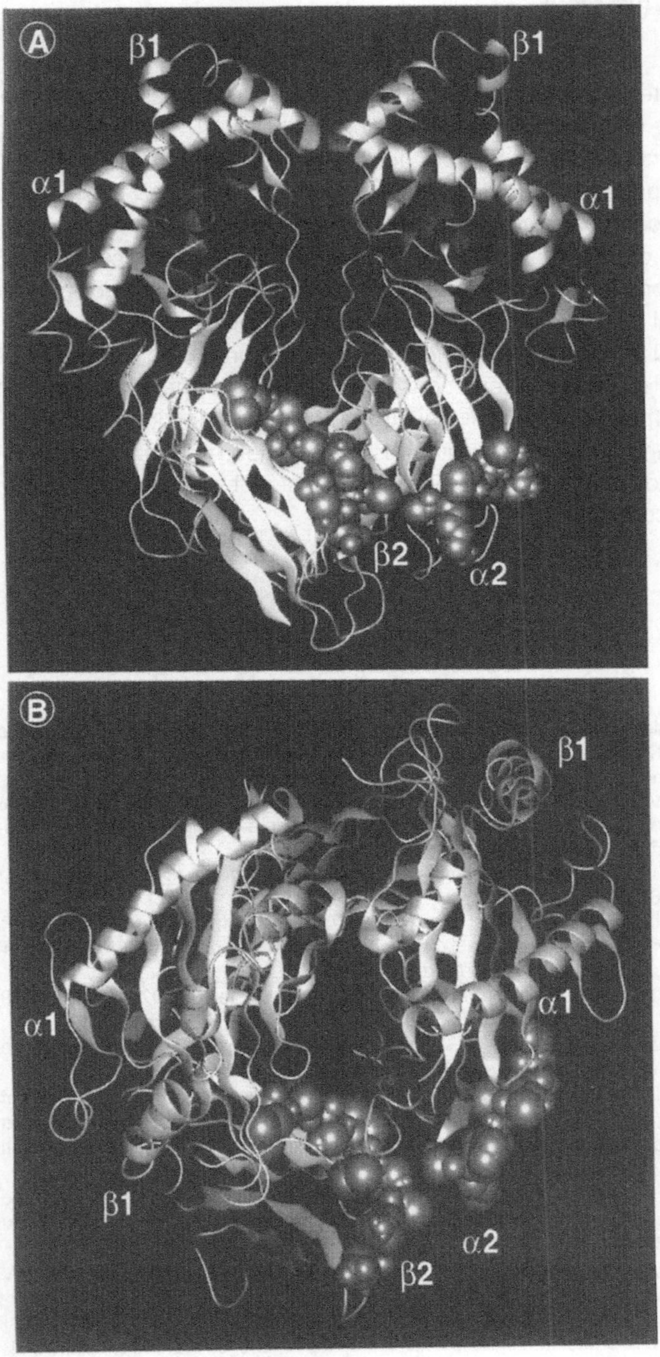


**Fig. 5A,B. A** Side view of a class II MHC molecule superdimer (ribbon drawing), with the α2 and β2 regions affecting CD4 interaction shown in *dark gray* atomic outline. **B** Top view of a class II MHC molecule superdimer (ribbon drawing), with the α2 and β2 regions affecting CD4 interaction shown in *dark gray* atomic outline. *α1, α2, β1, β2,* domains of the MHC class II molecule

## 8 Concluding Remarks

CD4 initially constituted a marker for identifying T cells with particular functional abilities. Recognition of the ability of CD4 to bind to class II molecules led to investigation of the structural basis for this key molecular interaction in the initiation of T cell responses. The data emerging from these studies, which have proved complex and controversial, have recently begun to point the way to a new understanding of the organization of active TCR-containing signaling complexes, providing evidence for a key role of the CD4 coreceptor in the oligomerization of receptor-based signal transduction units necessary for effective second messenger accumulation and the downstream induction of effector function. The further dissection of CD4-MHC class II interactions and the detailed characterization of the architecture of the oligomeric signaling complex will no doubt prove of fundamental importance to understanding of how antigen recognition is translated into T cell effector activity and how this process might be regulated by subtle changes in ligand recognition.

## References


Anderson P, Blue ML, Schlossman SF (1988) Comodulation of CD3 and CD4. Evidence for a specific association between CD4 and approximately 5% of the CD3: T cell receptor complexes on helper T lymphocytes. J Immunol 140: 1732–1737

Ashkenazi A, Presta LG, Marsters SA, Camerato TR, Rosenthal KA, Fendly BM, Capon DJ (1990) Mapping the CD4 binding site for human immunodeficiency virus by alanine-scanning mutagenesis. Proc Natl Acad Sci USA 87: 7150–7154

Ashwell JD, Klausner RD (1990) Genetic and mutational analysis of the T-cell antigen receptor. Annu Rev Immunol 8: 139–167

Biddison WE, Rao PE, Talle MA, Goldstein G, Shaw S (1982) Possible involvement of the OKT4 molecule in T cell recognition of class II HLA antigens. Evidence from studies of cytotoxic T lymphocytes specific for SB antigens. J Exp Med 156: 1065–1076

Bierer BE, Sleckman BP, Ratnofsky SE, Burakoff SJ (1989) The biologic roles of CD2, CD4, and CD8 in T-Cell activation. Annu Rev Immunol 7: 579–599

Bjorkman PJ, Saper MA, Samraoui B, Bennett WS, Strominger JL, Wiley DC (1987) Structure of the human class I histocompatibility antigen, HLA-A2. Nature 329: 506–512

Bowman MR, MacFerrin KD, Schreiber SL, Burakoff SJ (1990) Identification and structural analysis of residues in the V1 region of CD4 involved in interaction with human immunodeficiency virus envelope glycoprotein gp120 and class II major histocompatibility complex molecules. Proc Natl Acad Sci USA, 87: 9052–9056

Braunstein NS, Germain RN (1987) Allele-specific control of Ia molecule surface expression and conformation: implications for a general model of Ia structure-function relationships. Proc Natl Acad Sci USA 84: 2921–2925

Brodsky MH, Warton M, Myers RM, Littman DR (1990) Analysis of the site in CD4 that binds to the HIV envelope glycoprotein, J Immunol 144: 3078–3086

Brown JH, Jardetzky TS, Gorga JC, Stern LJ, Urban RG, Strominger JL, Wiley DC (1993) Three-dimensional structure of the human class II histocompatibility antigen HLA-DR1. Nature 364: 33–39

Cammarota G, Scheirle A, Takacs B, Doran DM, Knorr R, Bannwarth W, Guardiola J, Sinigaglia F (1992) Identification of a CD4 binding site on the β2 domain of HLA-DR molecules. Nature, 356: 799–801

Clayton LK, Hussey RE, Steinbrich R, Ramachandran H, Husain Y, Reinherz EL (1988) Substitution of murine for human CD4 residues identifies amino acids critical for HIV-gp120 binding. Nature 335: 363–366

Clayton LK, Sieh M, Pious DA, Reinherz EL (1989) Identification of human CD4 residues affecting class II MHC versus HIV-1 gp120 binding. Nature 339: 548–551

Corr M, Slanetz AE, Boyd LF, Jelonek MT, Khilko S, Al-Ramadi BK, Kim YS, Maher SE, Bothwell ALM, Margulies DH (1994) T cell receptor/MHC class I- peptide interactions: affinity, kinetics, and specificity. Science 265: 946–949

Demotz S, Grey HM, Sette A (1990) The minimal number of class II MHC-antigen complexes needed for T cell activation. Science 249: 1028–1030

Dialynas DP, Wilde DB, Marrack P, Pierres A, Wall KA, Havran W, Otten G, Loken MR, Pierres M, Kappler J, Fitch FW (1983) Characterization of the murine antigenic determinant, designated L3T4a, recognized by monoclonal antibody GK1.5: expression of L3T4a by functional T cell clones appears to correlate primarily with class II MHC antigen-reactivity. Immunol Rev 74: 29–56

Dintzis HM, Dintzis RZ, Vogelstein B (1976) Molecular determinants of immunogenicity: the immunon model of immune response. Proc Natl Acad Sci USA 73: 3671–3675

Doyle C, Strominger JL (1987) Interaction between CD4 and class II MHC molecules mediates cell adhesion. Nature 330: 256–259

Dustin ML, Springer TA (1989) T-cell receptor cross-linking transiently stimulates adhesiveness through LFA-1. Nature 341: 619–624

Engleman EG, Benike CJ, Grumet FC, Evans RL (1981) Activation of human T lymphocyte subsets: helper and suppressor/cytotoxic T cells recognize and respond to distinct histocompatibility antigens. J Immunol 127: 2124–2129

Fitch FW (1986) T cell clones and T cell receptors. Microbiol. Rev 50: 50–69

Fleury S, Lamarre D, Meloche S, Ryu SE, Cantin C, Hendrickson WA, Sekaly RP (1991) Mutational analysis of the interaction between CD4 and class II MHC: class II antigens contact CD4 on a surface opposite the gp120-binding site. Cell 66: 1037–1049

Fleury S, Thibodean J, Croteau G, Labrecque N, Aronson H-E, Cantin F, Long EO, Sékaly R-P (1995) Polymorphism in the β2 domain of HLA-DR affects the interaction with CD4. J Exp Med 182 (in press)

Gay D, Maddon P, Sekaly R, Talle MA, Godfrey M, Long E, Goldstein G, Chess L, Axel R, Kappler J, Marrack P (1987) Functional interaction between human T-cell protein CD4 and the major histocompatibility complex HLA-DR antigen. Nature 328: 626–629

Gay D, Buus S, Pasternak J, Marrack P (1988) The T-cell accessory molecule CD4 recognizes a monomorphic determinant on isolated Ia. Proc Natl Acad Sci USA 85: 5629–5633

Germain RN (1994) MHC-dependent antigen processing and peptide presentation: providing ligands for T lymphocyte activation. Cell 76: 287–299

Golding H, McCluskey J, Munitz TI, Germain RN, Margulies DH, Singer A (1985) T-cell recognition of a chimaeric class II/class I MHC molecule and the role of L3T4. Nature 317: 425–427

Greenstein JL, Kappler J, Marrack P, Burakoff SJ (1984) The role of L3T4 in recognition of Ia by a cytotoxic, H-2D$^d$-specific T cell hybridoma. J Exp Med 159: 1213–1224

Harding CV, Unanue ER (1990) Quantitation of antigen-presenting cell MHC class II/peptide complexes necessary for T-cell stimulation. Nature 346: 574-576

Janeway CA Jr (1992) The T cell receptor as a multicomponent signaling machine: CD4/CD8 coreceptors and CD45 in T cell activation. Annu Rev Immunol 10: 645–674

Janeway CA Jr, Rojo J, Saizawa K, Dianzani U, Portoles P, Tite J, Haque S, Jones B (1989) The co-receptor function of murine CD4. Immunol Rev 109: 77–92

Julius M, Maroun CR, Haughn L (1993) Distinct roles for CD4 and CD8 as coreceptors in antigen receptor signaling. Immunol Today 14: 177–183

König R, Huang LY, Germain RN (1992) MHC class II interaction with CD4 mediated by a region analogous to the MHC class I binding site for CD8. Nature 356: 796–798

König R, Shen X, Germain RN (1995) Involvement of both MHC Class II α and β chains in CD4 function indicates a role for ordered oligomeritation in T Cell activation. J Exp Med 182 (in press)

Krensky AM, Clayberger C, Reiss CS, Strominger JL, Burakoff SJ (1982) Specificity of OKT4$^+$ cytotoxic T lymphocyte clones. J Immunol 129: 2001–2003

Kwong PD, Ryu SE, Hendrickson WA, Axel R, Sweet RM, Folena WG, Hensley P, Sweet RW (1990) Molecular characteristics of recombinant human CD4 as deduced from polymorphic crystals. Proc Natl Acad Sci USA 87: 6423–6427

Lamarre D, Ashkenazi A, Fleury S, Smith DH, Sekaly RP, Capon DJ (1989a) The MHC-binding and gp120-binding functions of CD4 are separable. Science 245: 743–746

Lamarre D, Capon DJ, Karp DR, Gregory T, Long EO, Sekaly RP (1989b) Class II MHC molecules and the HIV gp120 envelope protein interact with functionally distinct regions of the CD4 molecule. EMBO J 8: 3271–3277

Landau NR, Warton M, Littman DR (1988) The envelope glycoprotein of the human immunodeficiency virus binds to the immunoglobulin-like domian of CD4. Nature 334: 159–162

Lombardi G, Barber L, Aichinger G, Heaton T, Sidhu S, Batchelor JR, Lechler RI (1991) Structural analysis of anti-DR1 allorecognition by using DR1/H- 2E$^k$ hybrid molecules. Influence of the β2-domain correlates with CD4 dependence. J Immunol 147: 2034–2040

Marrack P, Endres R, Shimonkevitz R, Zlotnik A, Dialynas D, Fitch F, Kappler J (1993) The major histocompatibility complex-restricted antigen receptor on T cells. II. Role of the L3T4 Product. J Exp Med 158: 1077–1091

Matsui K, Boniface JJ, Reay PA, Schild H, Fazekas de St. Groth B, Davis MM (1991) Low affinity interaction of peptide-MHC complexes with T cell receptors. Science 254: 1788–1791

Mazerolles F, Amblard F, Lumbroso C, Lecomte O, Van-de-Moortele PF, Barbat C, Piatier-Tonneau D, Auffray C, Fischer A (1990) Regulation of T helper-B lymphocyte adhesion through CD4-HLA class II interaction. Eur J Immunol 20: 637–644

McCluskey J, Germain RN, Margulies DH (1985) Cell surface expression of an in vitro recombinant class II/class I major histocompatibility complex gene product. Cell 40: 247–257

McCluskey J, Munitz T, Boyd L, Germain RN, Coligan JE, Singer A, Margulies DH (1988) Cell surface expression of the amino-terminal domain of Aακ: Recognition of an isolated MHC antigenic structure by allospecific T cells but not alloantibodies. J Immunol 140: 2081–2089

Meuer SC, Hussey RE, Hodgdon JC, Hercend T, Schlossman SF, Reinherz EL (1982) Surface structures involved in target recognition by human cytotoxic T lymphocytes. Science 218: 471–473

Miceli MC, Parnes JR (1993) Role of CD4 and CD8 in T cell activation and differentiation. Adv Immunol 53: 59–122

Miceli MC, von Hoegen P, Parnes JR (1991) Adhesion versus coreceptor function of CD4 and CD8: role of the cytoplasmic tail in coreceptor activity. Proc Natl Acad Sci USA 88: 2623–2627

Mizukami T, Fuerst TR, Berger EA, Moss B (1988) Binding region for human immunodeficiency virus (HIV) and epitopes for HIV-blocking monoclonal antibodies of the CD4 molecule defined by site-directed mutagenesis. Proc Natl Acad Sci USA 85: 9273–9277

Moebius U, Clayton LK, Abraham S, Diener A, Yunis JJ, Harrison SC, Reinherz EL (1992a) Human immunodeficiency virus gp120 binding C'C" ridge of CD4 domain 1 is also involved in interaction with class II major histocompatibility complex molecules. Proc Natl Acad Sci USA 89: 12008–12012

Moebius U, Clayton LK, Abraham S, Harrison SC, Reinherz EL (1992b) The human immunodeficiency virus gp120 binding site on CD4: delineation by quantitative equilibrium and kinetic binding studies of mutants in conjunction with a high-resolution CD4 atomic structure. J Exp Med 176: 507–517

Moebius U, Pallai P, Harrison SC, Reinherz EL (1993) Delineation of an extended surface contact area on human CD4 involved in class II major histocompatibility complex binding. Proc Natl Acad Sci USA 90: 8259–8263

Nag B, Wada HG, Passmore D, Clark BR, Sharma SD, McConnell HM (1993) Purified beta-chain of MHC class II binds to CD4 molecules on transfected HeLa cells. J Immunol 150: 1358–1364

Okada M, Henney CS (1980) The differentiation of cytotoxic T cells in vitro. II. Amplifying factor(s) produced in primary mixed lymphocyte cultures against K/D stimuli require the presence of Lyt-2$^+$ cells but not lyt-1$^+$ cells. J Immunol 125: 300–307

Olshevsky U, Helseth E, Furman C, Li J, Haseltine W, Sodroski J (1990) Identification of individual human immunodeficiency virus type 1 gp120 amino acids important for CD4 receptor binding. J Virol 64: 5701–5707

Ottenhoff THM, Elferink DG, Hermans J, De Vries RRP (1985) HLA class II restriction repertoire of antigen-specific T cells. I. The main restriction determinants for antigen presentation are associated with HLA-D/DR and not with DP and DQ. Hum Immunol 13: 105–116

Parnes JR (1989) Molecular biology and function of CD4 and CD8. Adv Immunol 44: 265–31

Peterson A, Seed B (1988) Genetic analysis of monoclonal antibody and HIV binding sites on the human lymphocyte antigen CD4. Cell 54: 65–72

Potter TA, Rajan TV, Dick RF, Bluestone JA (1989) Substitution at residue 227 of H-2 class I molecules abrogates recognition by CD8-dependent, but not CD8-independent, Cytotoxic T lymphocytes. Nature 337: 73–75

Ramesh P, Barber L, Batchelor JR, Lechler RI (1992) Structural analysis of human anti-mouse H-2E xenorecognition: T cell receptor bias and impaired CD4 interaction contribute to weak xeno-responses. Int Immunol 4: 935–943

Ratcliffe MJ, Coggeshall KM, Newell MK, Julius MH (1992) T cell receptor aggregation, but not dimerization, induces increased cytosolic calcium concentrations and reveals a lack of stable association between CD4 and the T cell receptor. J Immunol 148: 1643–1651

Reinherz EL, Schlossman SF (1980) The differentiation and function of human T lymphocytes. Cell 19: 821–827

Reinherz EL, Kung PC, Goldstein G, Levey RH, Schlossman SF (1980) Discrete stages of human intrathymic differentiation: analysis of normal thymocytes and leukemic lymphoblasts of T-cell lineage. Proc Natl Acad Sci USA, 77: 1588–1592

Reinherz EL, Meuer SC, Schlossman SF (1983) The delineation of antigen receptors on human T lymphocytes. Immunol Today 4: 5–8

Rojo JM, Saizawa K, Janeway CA Jr (1989) Physical association of CD4 and the T cell receptor can be induced by anti-T cell receptor antibodies. Proc Natl Acad Sci USA 86: 3311–3315

Rosenstein Y, Burakoff SJ, Herrmann SH (1990) HIV-gp120 can block CD4-class II MHC-mediated adhesion. J Immunol, 144: 526–531

Rudd C, Helms S, Barber EK, Schlossman SF (1989) The CD4/CD8: P56lck complex in T lymphocytes: a potential mechanism to regulate T-cell growth. Biochem Cell Biol 67: 581–589

Ruppert J, Alexander J, Snoke K, Coggeshall M, Herbert E, McKenzie D, Grey HM, Sette A (1993) Effect of T-cell receptor antagonism on interaction between T cells and antigen-presenting cells and on T-cell signaling events. Proc Natl Acad Sci USA 90: 2671–2675

Ryu SE, Kwong PD, Truneh A, Porter TG, Arthos J, Rosenberg M, Dai XP, Xuong NH, Axel R, Sweet RW, Hendrickson WA (1990) Crystal structure of an HIV-binding recombinant fragment of human CD4. Nature, 348: 419–426

Saito T, Weiss A, Miller J, Norcross MA, Germain RN (1987) Specific antigen-Ia activation of transfected human T cells expressing murine Ti alpha beta-human T3 receptor complexes. Nature 325: 125–130

Saizawa K, Rojo J, Janeway CA Jr (1987) Evidence for a physical association of CD4 and the CD3: $\alpha$: $\beta$ T cell receptor. Nature 328: 260–263

Salter RD, Benjamin RJ, Wesley PK, Buxton SE, Garrett TPJ, Clayberger C, Krensky AM, Norment AM, Littman DR, Parham P (1990) A binding site for the T-cell coreceptor CD8 on the $\alpha$3 domain of HLA-A2. Nature 345: 41–46

Seed B, Aruffo A (1987) Molecular cloning of the CD2 antigen, the T-cell erythrocyte receptor, by a rapid immunoselection procedure. Proc Natl Acad Sci USA 84: 3365–3369

Spits H, Borst J, Terhorst C, de Vries JE (1982) The role of T cell differentiation markers in antigen-specific and lectin-dependent cellular cytotoxicity mediated by $T8^+$ and $T4^+$ human cytotoxic T cell clones directed at class I and class II MHC antigens. J. Immunol., 129: 1563–1569

Swain SL (1981) Significance of LyT phenotypes: Lyt-2 antibodies block activities of T cells that recognize class I major histocompatibility complex antigens regardless of their function. Proc Natl Acad Sci U.S.A., 78: 7101–7105

Swain SL (1983) T cell subsets and the recognition of MHC class. Immunol Rev 74: 129–142

Swain SL, Dutton RW, Schwab R, Yamamoto J (1983) Xenogeneic human anti-mouse T cell responses are due to the activity of the same functional T cell subsets responsible for allospecific and major histocompatibility complex-restricted responses. J Exp Med 157: 720–729

Swain SL, Dialynas D, Fitch FW, English MJ (1984) Monoclonal antibody to L3T4 blocks the function of T cells specific for class II major histocompatibility complex antigens. J Immunol 132: 1118–1123

Symer DE, Dintzis RZ, Diamond DJ, Dintzis HM (1992) Inhibition or activation of human T cell receptor transfectants is controlled by defined, soluble antigen arrays. J Exp Med 176: 1421–1430

Trowsdale J, Ragoussis J, Campbell RD, Holborn LU (1991) Map of the human MHC. Immunol Today 12: 443–446

Ullrich A, Schlessinger J (1990) Signal transduction by receptors with tyrosine kinase activity. Cell 61: 203–212

Veillette A, Bookman MA, Horak EM, Bolen JB (1988) The CD4 and CD8 T cell surface antigens are associated with the internal membrane tyrosine-protein kinase $p56^{lck}$. Cell 55: 301–308

Veillette A, Bookman MA, Horak EM, Samelson LE, Bolen JB (1989) Signal transduction through the CD4 receptor involves the activation of the internal membrane tyrosine-protein kinase $p56^{lck}$ Nature 338: 257–259

Vignali DA, Moreno J, Schiller D, Hämmerling GJ (1992) Species-specific binding of CD4 to the $\beta$2 domain of major histocompatibility complex class II molecules. J Exp Med 175: 925–932

Vignali DA, Doyle C, Kinch MS, Shin J, Strominger JL (1993) Interactions of CD4 with MHC class II molecules, T cell receptors and $p56^{lck}$. Philos. Trans R Soc Lond [Biol] 342: 13–24

von Boehmer H (1988) The developmental biology of T lymphocytes. Annu Rev Immunol 6: 309–326
von Boehmer H (1994) Positive selection of lymphocytes. Cell 76: 219–228
von Hoegen P, Miceli MC, Tourvieille B, Schilham M, Parnes JR (1989) Equivalence of human and mouse CD4 in enhancing antigen responses by a mouse class II-restricted T cell hybriodoma. J Exp Med 170: 1879–1886
Wang JH, Yan YW, Garrett TP, Liu JH, Rodgers DW, Garlick RL, Tarr GE, Husain Y, Reinherz EL, Harrison SC (1990) Atomic structure of a fragment of human CD4 containing two immunoglobulin-like domains. Nature 348: 411–418
Weber S, Karjalainen K (1993) Mouse CD4 binds MHC class II with extremely low affinity. Int Immunol 5: 695–698
Weber S, Traunecker A, Oliveri F, Gerhard W, Karjalainen K (1992) Specific low-affinity recogition of major histocompatibility complex plus peptide by soluble T-cell receptor. Nature 356: 793–796
Wilde DB, Marrack P, Kappler J, Dialynas DP, Fitch FW (1983) Evidence implicating L3T4 in class II MHC antigen-reactivity: monoclonal antibody GK1.5 (anti-L3T4a) blocks class II MHC antigen-specific proliferation, release of lymphokines, and binding by cloned murine helper T lymphocyte lines. J Immunol 131: 2178–2183
Xu H, Littman DR (1993) A kinase-independent function of Lck in potentiating antigen-specific T cell activation. Cell 74: 633–643

# CD4 and Signal Transduction

K.S. RAVICHANDRAN[1], T.L. COLLINS[1], and S.J. BURAKOFF[1,2]



## 1 Role of CD4 in Cell-Cell Adhesion and Signaling

CD4 is a 55–60 kDa monomeric glycoprotein expressed on approximately 60% of peripheral blood T lymphocytes (REINHERZ et al. 1979; TERHORST et al. 1980). Originally identified as a phenotypic marker (REINHERZ et al. 1979; WEBB et al. 1979; DIALYNAS et al. 1983), a role for CD4 in the activation of T lymphocytes was indicated when anti-CD4 antibodies inhibited a variety of T lymphocyte effector functions in vitro (KRENSKY et al. 1982; SPITS et al. 1982; WILDE et al. 1983; ROGOZINSKI et al. 1984). It was later demonstrated that CD4 facilitates T lymphocyte function by binding to the nonpolymorphic region of the major histocompatibility (MHC) class II antigens expressed on the surface of antigen presenting cells (APCs) (DOYLE and STROMINGER 1987; SLECKMAN et al. 1987; GAY et al. 1988;

[1]Division of Pediatric Oncology, Dana-Farber Cancer Institute, 44 Binney Street, Boston, MA 02115, USA
[2]Department of Pediatrics, Harvard Medical School, 200 Longwood Avenue, Boston, MA 02115, USA

Cammarota et al. 1992; König et al. 1992). The interaction between CD4 and MHC class II molecules have proven to be critical for antigen responsiveness of CD4⁺T cells (Gay et al. 1987; Sleckman et al. 1987).

## 1.1 Evidence for CD4 as an Adhesion Molecule

Although these studies provided crucial insights on CD4 function in T cell function, it was not clear whether CD4 served primarily to increase adhesion between the T lymphocyte and the APC, to transmit additional signals during T lymphocyte engagement, or both. Anti-CD4 antibodies inhibited cytolysis of target cells by preventing conjugate formation, suggesting that CD4 may play a role in T cell adhesion (Biddison et al. 1984). The ability of anti-CD4 antibodies to inhibit conjugate formation was inversely related to the avidity of the T cell/APC interaction (Biddison et al. 1984), suggesting that, when the avidity of the T cell receptor (TCR) for antigen/MHC is low, the CD4:MHC class II interaction may become critical. Similar results were seen using a T-helper hybridoma which produced interleukin-2 (IL-2) upon stimulation with antigen plus MHC class II-bearing cells (Marrack et al. 1983; Greenstein et al. 1985). Taken together, these data suggest that CD4 increases the avidity of T cells for MHC class II-bearing cells, particularly in situations when the antigen is limiting or when the avidity of the TCR for antigen/MHC is low.

## 1.2 Evidence for CD4 as a Signaling Molecule

Several observations suggested that CD4 can transduce intracellular signals during T cell activation. Cross-linking CD4 with the TCR/CD3 complex caused significantly greater activation than if the TCR/CD3 complex was cross-linked alone (Anderson et al. 1987; Eichmann et al. 1987; Emmrich et al. 1986, 1987; Owens et al. 1987; Walker et al. 1987). These effects were most dramatic at suboptimal concentrations of the anti-TCR or anti-CD3 antibody (Eichmann et al. 1987), a situation analogous to low antigen concentrations or low avidity of TCR for MHC-class II in which CD4 signaling may be most critical. It is important to note that in these assays CD4 was physically joined to the TCR/CD3 complex, since in several of these reports it was also demonstrated that ligation of CD4 separately from the TCR/CD3 complex could inhibit T cell activation (Anderson et al. 1987; Emmrich et al. 1987; Walker et al. 1987). This also provided an explanation for the initial findings that some anti-CD4 antibodies had a negative effect on T lymphocyte activation (Bank and Chess 1985; Wassmer et al. 1985). Thus the negative effects of CD4 cross-linking could be due to the sequestration of CD4 away from the TCR/CD3 complex rather than due to the generation of a negative signal by CD4.

The ability of CD4 to actively participate in signal transduction was emphasized by the work of Sleckman et al. (1988), who demonstrated that the enhancement of T lymphocyte activation resulting from expression of CD4 could be prevented by deletion of the cytoplasmic domain of CD4, or by replacement of the cytoplasmic and transmembrane domains with a glycophosphatidylinositol anchor (SLECKMAN et al. 1991). Since the glycophosphatidylinositol-linked CD4 mutant had the same avidity for MHC class II protein as the full-length CD4 (SLECKMAN et al. 1991), this indicated that the enhancement of T cell activation by ligation of CD4 with MHC class II molecules was dependent on the cytoplasmic domain of CD4 and not merely on the extracellular, ligand-binding domain.

These findings, however, do not exclude a role for CD4 in cell-cell adhesion. In fact, it was shown that the loss of CD4-mediated enhancement of T lymphocyte activation resulting from deletion of the cytoplasmic tail of CD4 can be overcome by expression of very high amounts of the CD4 deletion mutant (SLECKMAN et al. 1988). These assays were done in vitro using a murine T cell hybridoma; however, the same results have recently been demonstrated in vivo (KILLEEN and LITTMAN 1993). A cytoplasmic tail-deletion mutant of CD4, expressed as a transgene in mice lacking CD4, could rescue the functional development of CD4$^+$T-helper lymphocytes when overexpressed but not at physiological levels (KILLEEN and LITTMAN 1993). These data support the in vitro observations, implicating CD4 in both adhesion and signaling in immature and mature T cells.

## 1.3 Interaction of CD4 with the T Cell Receptor/CD3 Complex

A number of observations suggest that CD4 associates with the TCR/CD3 complex during T cell activation and that this association may be critical for CD4 mediated enhancement of T lymphocyte responsiveness. CD4 redistributes with the TCR/CD3 complex to the site of contact between the T cell and an APC-bearing specific antigen (antigen which binds the TCR) but does not redistribute to the site of contact with an APC-bearing nonspecific antigen (KUPFER et al. 1987). This indicates that CD4 redistribution is dependent on TCR recognition. CD4 can be induced to comodulate with the TCR following stimulation with anti-TCR antibodies (SAIZAWA et al. 1987); similarly, anti-TCR antibodies can induce CD4 to cocap with the TCR/CD3 complex under activating conditions (ROJO et al. 1989). Interestingly, the ability of anti-TCR antibodies to cause CD4 cocapping correlates with the efficiency of antibody mediated activation of the T lymphocyte (PORTOLES et al. 1989; ROJO et al. 1989; DIANZANI et al. 1992). Anti-TCR antibodies which induce CD4 cocapping with the TCR/CD3 complex also cause phosphorylation of the TCR-ζchain much better than antibodies which do not induce CD4 cocapping (DIANZANI et al. 1992). These data suggest that the coassociation of CD4 with the TCR/CD3 complex has important consequences for T cell activation.

A method to quantitate the CD4 coassociation with the TCR/CD3 complex was developed by MITTLER et al. (1989), which allows an analysis of the requirements for CD4-TCR/CD3 coassociation. Using fluorescence resonance energy transfer (FRET), the authors demonstrated that CD4 rapidly coassociates with the TCR/CD3 complex following activation via antibody cross-linking of CD3ε (maximum coassociation is achieved within 10 min at 37°C). This coassociation is completely dependent on the expression of the cytoplasmic domain of CD4, since CD4 deletion mutants failed to coassociate with the TCR/CD3 complex (Mittler et al. 1989). The coassociation of CD4 with the TCR/CD3 complex is specifically dependent on the association of the protein tyrosine kinase Lck with CD4 (Collins et al. 1992 and see below).

## 1.4 Interaction of CD4 and Lck

CD4 associates noncovalently with the Src-related protein tyrosine kinase Lck (Rudd et al. 1988; Veillette et al. 1988; Barber et al. 1989) suggesting that CD4 signaling may be mediated, at least in part, by the activation of Lck. Lck is a 56 kDa protein composed of a unique NH2-terminal domain, a src-homology 2 (SH2) domain, a src-homology 3 (SH3) domain and a kinase domain. The CD4: Lck interaction occurs between the cytoplasmic domain of CD4 (Shaw et al. 1989, 1990; Vega et al. 1990; Veillette et al. 1990) and the NH2-terminal domain of Lck (Shaw et al. 1989). It has been shown that cysteine residues 420 and 422 of human CD4 and residues 20 and 23 of Lck (Turner et al. 1990) are critical for this interaction. Cross-linking of CD4 with monoclonal antibodies (mAbs) results in a rapid increase in the phosphorylation of Lck on both tyrosine and serine residues, an increase in Lck activity (Marth et al. 1989; Luo and Sefton 1990) and an increase in tyrosine phosphorylation of a number of cellular proteins (Veillette et al. 1989 a, b). These data suggested that Lck may be involved in CD4-mediated signal transduction.

## 1.5 CD4/Lck Association Is Necessary for Coreceptor Function via CD4

Mutation of either of the two cysteine residues in the cytoplasmic domain of CD4 involved in Lck binding results in a loss of association between CD4 and Lck (Turner et al. 1990). Whereas expression of wild-type CD4 in a murine T cell hybridoma results in vastly enhanced antigen responsiveness (Gay et al. 1987; Sleckman et al. 1987; Glaichenhaus et al. 1991), expression of CD4 point mutants (in which the cysteines necessary for interaction with Lck have been mutated) did not enhance antigen responsiveness (Glaichenhaus et al. 1991; Collins et al. 1992). Interestingly, these CD4 mutants also failed to coassociate with the TCR/CD3 complex during T cell stimulation (Collins et al. 1992). Since IL-2 production in response to cross-linking with anti-TCR and anti-CD4 antibodies was also diminished (Collins et al. 1992), these data suggest that the loss in CD4 enhancement of antigen responsiveness is due largely to a lack of CD4/Lck mediated signaling.

## 1.6 CD4/Lck and Ras Signaling

One of the early signaling events following TCR triggering, which is downstream of tyrosine kinases, is the activation of Ras proteins. Products of *ras* oncogenes play a crucial role in normal cellular mitogenesis (DOWNWARD 1992). Ras proteins are small guanine nucleotide-binding proteins of 21 kDa and signaling via Ras is determined by the type of nucleotide bound to Ras. Guanine triphosphate (GTP)-bound Ras is active while guanine diphosphate (GDP) bound Ras is inactive. The nucleotide status of Ras is tightly regulated by the nucleotide exchange factors which load GTP for GDP onto Ras and by GTPase activating proteins (GAPs) which augment the intrinsic GTPase activity of Ras proteins. In fibroblasts, binding of a number of peptide growth factors to their receptors leads to conversion of Ras to its active GTP-bound state. Signaling through these receptors, which are themselves tyrosine kinases, is blocked by mutational inactivation of *ras* genes or dominant negative inhibitors of Ras proteins. The activation of the receptor-associated tyrosine kinase activity appears to be crucial for Ras activation, indicating that the tyrosine kinases are upstream of Ras (reviewed in SCHLESSINGER 1993). In T cells, studies using constitutively active or dominant-negative Ras proteins have demonstrated that Ras activation is important in TCR-mediated transcription initiation of the IL-2 gene (BALDARI et al. 1992; RAYTER et al. 1992). Moreover, in cells in which transcription from the IL-2 promoter has been initiated by expression of a constitutively active form of Lck, introduction of a dominant negative form of Ras inhibited transcription from the IL-2 promoter, suggesting that Ras functions downstream of Lck in T cells.

## 1.7 CD4/Lck and the Involvement of Shc and Vav in Ras Signaling

Two independent regulatory pathways of Ras activation has been described in T cells: one which is downstream of tyrosine kinase activation and mediated via the TCR and another, which is unique to lymphocytes, is protein kinase C (PKC)-dependent and is apparently tyrosine kinase-independent (IZQUIERDO et al. 1992). While the pathway involved in PKC-mediated Ras activation remains unclear, several intracellular signaling proteins have been shown to be involved in the tyrosine kinase mediated Ras activation (Fig. 1).

One of these proteins, Shc, has been shown to be upstream of Ras in neurofibroma cells and is a good substrate for Src, Fps and Lck tyrosine kinases in fibroblasts (MCGLADE et al. 1992; ROZAKIS-ADCOCK et al. 1992). Shc is one of the adapter proteins, composed of a single SH2 domain, a collagen-homology domain and a unique NH2-terminal domain, but Shc lacks any apparent catalytic activity (PELICCI et al. 1992). Upon TCR activation Shc is tyrosine-phosphorylated and it subsequently interacts with another adapter protein Grb2 (RAVICHANDRAN et al. 1993). Grb2 also lacks a catalytic domain and is composed of one SH2 domain and two SH3 domains (LOWENSTEIN et al. 1992). The SH2 and SH3 domains, found in a number of signaling proteins, mediate specific protein-protein interactions. While SH2 domains bind phosphotyrosine-containing sequences, SH3 domains bind proline-rich sequences (REN et al. 1993; SONGYANG et al. 1993). Grb2, via its SH2 domain, interacts with phosphorylated Shc and also interacts via its SH3 domains

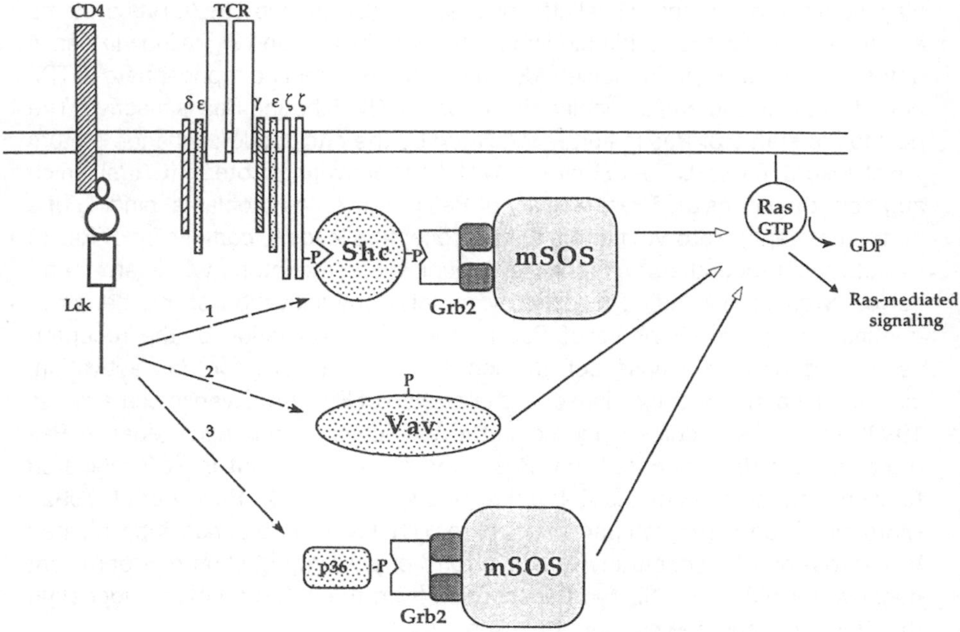


**Fig. 1.** CD4/Lck may influence T cell receptor (*TCR*)-mediated Ras activation in at least three different ways: (1) Tyrosine phosphorylation of Shc is augmented by cross-linking of CD4 with the TCR and, in turn, the level of Grb2 bound to Shc. Since the interaction of phosphorylated Shc with Grb2 regulates the Grb2: mSOS association, this may in turn lead to enhanced Ras activation; (2) Vav has been shown to directly act as a guanine nucleotide exchange factor for Ras and tyrosine phosphorylation of Vav (mediated by Lck) influences its exchange activity toward Ras; and (3) Phosphorylation of the yet to be identified p36 protein leads to its interaction with Grb2 and mSOS. Whether these are three independent mechanisms of Ras activation that occur concurrently or used at different stages of T cell activation remains to be determined

with mSOS, a GTP/CDP exchange factor for Ras (Buday and Downward 1993; Egan et al. 1993). Shc, via its own SH2 domain, also interacts with the tyrosine-phosphorylated TCR-ζ chain. Thus, Shc and Grb2 acting as adapter proteins help to bring mSOS to the membrane, thereby leading to Ras activation. Another unidentified tyrosine-phosphorylated protein, p36, has also been shown to interact with Grb2 and mSOS upon T cell activation (Buday et al. 1994; Sieh et al. 1994).

While TCR clustering alone leads to Shc phosphorylation, cross-linking of CD4 together with the TCR greatly enhances Shc phosphorylation and also increases the level of Grb2 and mSOS bound to Shc (Ravichandran et al. 1993). This suggests that CD4/Lck, through phosphorylation of Shc, may play a role in regulating the level of Ras activation. Since the interaction of tyrosine-phosphorylated Shc with the Grb2-SH2 domain seems to enhance the interaction of mSOS with the Grb2-SH3 domains (Ravichandran et al. 1995), the enhanced phosphorylation of Shc mediated by CD4/Lck may further augment TCR-dependent Ras activation. That Lck may be the kinase which phosphorylates Shc during

T cell activation is supported by several observations. First, Shc is a very good substrate for Lck in vitro, and in vitro phosphorylated Shc binds Grb2 and mSOS, suggesting that phosphorylation of Shc by Lck occurs on functionally relevant sites (RAVICHANDRAN et al. 1995). Second, transient expression of mutant forms of Shc (in which the Grb2 binding site has been mutated) was able to inhibit the transcription initiated from the IL-2 promoter by a constitutively active form of Lck (Pelicci et al., personal communication). These data suggest that Lck may phosphorylate Shc in vivo, but do not rule out a role for other kinases. The level of p36 phosphorylation is also enhanced by CD4 cross-linking with the TCR, and how it influences its interaction with Grb2 and mSOS remains to be determined.

Recently, the SH2 and SH3 domain-containing protein Vav has been shown to be capable of acting as a guanine nucleotide exchange factor for Ras following TCR activation (GULBINS et al. 1993). It has also been shown that the *dbl* homology domain of Vav is capable of mediating the nucleotide exchange activity. Phosphorylation of Vav by Lck in vitro was sufficient to induce its nucleotide exchange activity toward Ras. Moreover, Vav also has a region which binds to diacylglycerol and this also appears to regulate its nucleotide exchange activity on Ras. The individual contributions of Shc, p36 or Vav to the overall formation of GTP-bound Ras following TCR activation is currently unknown. It is likely that different mechanisms of Ras activation are used under different activation conditions or at different stages of T cell maturation. The extent to which CD4/Lck influences tyrosine kinase mediated Ras activation in T cells awaits further investigation.

In addition to its putative positive role in Ras activation, Lck may also play a negative role through its interaction with the p120 Ras-GAP. GAP is a GTPase activating protein which enhances the intrinsic GTPase activity of Ras by more than 10 000-fold. Using recombinant proteins, it was shown that GAP is a substrate for Lck in vitro and that GAP interacts with Lck. The interaction of GAP with Lck also led to an enhancement of the tyrosine kinase activity of Lck in vitro (AMREIN et al. 1992). Since GAP is a complex protein with multiple domains and interacts with at least two other proteins besides Lck, elucidation of the role for GAP during T cell activation and the interplay between GAP and Lck requires further experiments.

## 2 Role of CD4/Lck in Lipid Messenger Signaling

### 2.1 Lck and Phosphoinositide Signaling

One of the earliest events following TCR triggering is the tyrosine phosphorylation of phospholipase Cγ1 (PLCγ1), an enzyme which cleaves phosphoinositol into inositol tris-phosphate (IP3) and diacylglycerol (PARK et al. 1991; WEISS et al. 1991). Production of IP3 leads to release of calcium from intracellular stores while diacylglycerol binds to and activates the serine/threonine kinase PKC. Association of PLCγ1 with platelet-derived growth factor (PDGF) receptor and the subsequent tyrosine phosphorylation of PLCγ1 appears to be essential for mitogenesis mediated by PDGF in fibroblasts (VALIUS and KAZLAUSKAS 1993). Although no association between PLCγ1 and TCR has been shown to date, the association between tyrosine-phosphorylated Lck and the SH2 domain of PLCγ1 has been documented (WEBER et al. 1992). It is possible that Lck phosphorylation may serve to recruit PLCγ1 to the membrane during T cell activation and that Lck may subsequently phosphorylate PLCγ1. It is noteworthy that the PLCγ1 interaction with a membrane-associated 36 kDa protein has also been demonstrated (SIEH et al. 1994) and may also serve as a mechanism for PLCγ1 recruitment.

## 2.2 Lck and Lipid Kinases

Recently it has been demonstrated that CD4/Lck can associate with two lipid kinases: phosphoinositide-3 (PI-3) kinase and phosphoinositide-4 (PI-4) kinase (PRASAD et al. 1993a, b; KAPELLER et al. 1994). While PI-3 kinase activity seems to associate with CD4 via Lck, PI-4 kinase associates directly with CD4. The association of PI-3 kinase with the PDGF receptor is essential for PDGF-mediated mitogenesis in fibroblasts (VALIUS and KAZLAUSKAS 1993); by analogy, PI-3 kinase would be expected to play an important role in T cell activation. PI-3 kinase is composed of a 110 kDa catalytic subunit (p110) and an 85 kDa regulatory subunit (p85). p85, which contains two SH2 domains, an SH3 domain and two proline-rich SH3 domain-binding 'sequences, helps to recruit p110 to the membrane. The association of p85 with PDGF receptor, the insulin-receptor substrate-1 and epidermal growth factor receptor is mediated by the interaction of the p85 SH2 domain with the tyrosine-phosphorylated receptor (PAWSON and GISH 1992). However, the interaction of Lck with p85 is mediated by binding of the Lck-SH3 domain with the proline-rich sequences in p85. Although there is a low level of PI-3 kinase activity constitutively associated with CD4/Lck, the level of PI-3 kinase activity associated with CD4/Lck is enhanced upon cross-linking of CD4 either with anti-CD4 antibody or HIV gp120 (PRASAD et al. 1993b). It is noteworthy that PI-3 kinase also associates with the Fyn tyrosine kinase (a similar interaction as with Lck) (PRASAD et al. 1993a; KAPELLER et al. 1994) and the T cell costimulatory molecule CD28 (STEIN et al. 1994; TRUITT et al. 1994). The relative significance of the CD28, Fyn and CD4/Lck mediated recruitment of PI-3 kinase to the TCR/CD3 complex during T cell activation remains to be determined.

## 2.3 Interplay Between Lck and Other Tyrosine Kinases and Phosphatases

It has been shown that both serine-threonine phosphatases and tyrosine phosphatases can be regulated via tyrosine phosphorylation. There is evidence that Lck can phosphorylate PP2a, a serine-threonine phosphatase (Chen et al. 1992), and SHPTP1, a tyrosine phosphatase expressed predominantly in hematopoietic cells (Lorenz et al. 1994). While the enzymatic activity of PP2a is enhanced following Lck mediated phosphorylation, the effect of tyrosine phosphorylation on SHPTP1 has been technically difficult to assess due to the autodephosphorylation by SHPTP1. Since mice deficient in SHPTP1 phosphatase activity have profound alterations in their hematopoietic development, a very short lifespan, and are reported to carry T cell defects, the enzymatic activity of SHPTP1 and its regulation may be essential for normal signaling (Shultz et al. 1993). Cross-linking of CD4 (either alone or in conjunction with TCR) on thymocytes and mature T cells leads to tyrosine phosphorylation of SHPTP1. It is tempting to speculate that SHPTP1 activity may be regulated by Lck and that SHPTP1 may act on critical substrate(s) during T cell signaling.

Recently, it has been demonstrated that Lck, via its SH2 domain, can interact with phosphorylated ZAP-70 (Duplay et al. 1994). Given the evidence that Lck functions upstream of ZAP-70 (Iwashima et al. 1994 and see below), the possibility exists that Lck could be recruited at different stages of T cell activation (see below).

## 3 Paradoxes in CD4/Lck Mediated Signaling

Tyrosine phosphorylation of cytoplasmic proteins is one of the earliest events following T cell receptor ligation (Hsi et al. 1989). Activation of protein tyrosine kinases precedes the release of intracellular calcium. Since tyrosine kinase inhibitors abrogate TCR-mediated signaling (June et al. 1990a,b), the tyrosine kinase activity seems to be essential for the activation of T lymphocytes. Cross-linking of CD4 results in rapid activation of Lck protein tyrosine kinase activity (Veillette et al. 1989a,b), and therefore it has been proposed that the enhancement of T cell activation which results from the ligation of CD4 during antigen response is due to the kinase activity of Lck.

Although the precise temporal order for the recruitment and activation of different tyrosine kinases following TCR triggering is not fully determined, the available data suggest that Lck of Fyn may initiate the earliest tyrosine phosphorylation events leading to phosphorylation of the TCR-$\zeta$ chain. The ZAP-70 tyrosine kinase is also recruited to the membrane (through its interaction with phosphorylated TCR-$\zeta$ and $\varepsilon$ chains) (Straus and Weiss 1992; Wange et al. 1993), which, in turn, may amplify the signal. The finding that TCR-$\zeta$ must be phosphorylated on tyrosine for ZAP-70 to associate suggests that ZAP-70 does not initiate the tyrosine kinase cascade. Transient transfections in COS cells expressing Lck, ZAP-70 and $\zeta$ support this model (Iwashima et al. 1994). The profound defect in tyrosine phosphorylation of intracellular substrates in cells lacking Lck kinase activity strongly implicates Lck as one of the earliest kinases involved in T cell signaling.

## 3.1 Tyrosine Kinase Activity of Lck May Not Be Necessary for CD4 Function as a Coreceptor

It was startling to discover that CD4-associated tyrosine kinase activity of Lck could be reduced or abrogated without a loss in CD4-dependent T lymphocyte activation (Xu and Littman 1993; Collins and Burakoff 1993). Expression of a chimeric receptor containing CD4 extracellular and transmembrane domains fused to a kinase-deleted Lck resulted in similar IL-2 production following stimulation with CD4-dependent antigen as when a full-length, kinase-functional form of Lck was fused to CD4 (Xu and Littman 1993). Furthermore, over- expression of a deletion mutant of Lck, lacking the kinase domain, resulted in a dramatic inhibition of tyrosine kinase activity precipitable with CD4 but did not diminish IL-2 production in response to CD4-dependent antigen or to cross-linking with anti-CD4 and anti-TCR antibodies (Collins and Burakoff 1993). These data suggest that the kinase domain of Lck may not be necessary for CD4 enhancement of T lymphocyte activation. In addition to the kinase domain, Lck also contains an SH2 domain and an SH3 domain. It is possible that CD4 coreceptor activity is mediated by either, or both, of these domains.

## 3.2 Does the CD4-Associated Lck Function More Like an Adapter Protein Than a Kinase?

It has been difficult to reconcile these observations with the finding that tyrosine kinase activity of Lck is crucial for signaling in mature T cells (observed in cells lacking Lck kinase activity) (Karnitz et al. 1992; Straus and Weiss 1992) and for the necessity of Lck for thymic development (observed in Lck-deficient mice) (Molina et al. 1992). Several possibilities, which are not mutually exclusive, may explain these observations. First, Lck may mediate TCR/CD3 signals in a CD4-independent fashion. This is supported by the finding that a constitutively active form of Lck can substitute for CD4 in enhancing antigen responsiveness (Abraham et al. 1991). Second, the kinase activity of Lck is not necessary if other Src-family kinases are expressed at sufficient levels. Overexpression of Fyn or expression of a constitutively active form of Fyn can enhance antigen responsiveness similar to a constitutively active form of Lck (Cooke et al. 1991; Davidson et al. 1992). Moreover, expression of a constitutively active form of Src, which is not normally expressed in T cells, results in phosphorylation of TCR components and spontaneous IL-2 production (O'Shea et al. 1991). Therefore, it is possible that the tyrosine phosphorylation of TCR-ζ may be achieved by any available Src-family tyrosine kinase and does not specifically require Lck kinase activity.

The third possibility is that the low level of Fyn associated with the TCR/CD3 complex (Samelson et al. 1990) or free Lck (not associated with CD4) might initiate phosphorylation of CD3 chains, (especially ζ and ε) to begin recruiting ZAP-70. At this time, CD4 associated Lck may be recruited to the TCR/CD3 complex either directly (via its SH2 domain interacting with the phosphorylated TCR-ζ or CD3 chains) or through its interaction with phosphorylated ZAP-70 (DUPLAY et al. 1994). Concurrently, Lck may bring other proteins into the TCR/CD3 complex (via its SH2 and SH3 domains) which are essential for CD4/Lck mediated coreceptor activity (Fig. 2).

There are several observations that seem to support this model. In both reports mentioned above (COLLINS and BURAKOFF 1993; XU and LITTMAN 1993), in which CD4/Lck kinase activity was found to be unnecessary for CD4 coreceptor function, the cells used also expressed normal levels of endogenous wild-type Lck. Thus, it cannot be ruled out that the endogenous Lck played a role in initiating the T cell response. Also both of the studies used the same functional read-out, i.e., IL-2 production, to assess the necessity of CD4-associated Lck kinase activity. If the CD4:Lck kinase activity were required for other functions during early stages of T cell signaling, they may have been missed by these studies. However, it was found that the SH2 domain of the kinase-deficient Lck was essential for the CD4 coreceptor activity. This would suggest that the CD4-associated Lck functions more like an adapter protein, possibly bringing additional proteins into the TCR/CD3 complex via its SH2 and SH3 domains. The association of PI-3 kinase activity with CD4 (via Lck) supports this notion of CD4-associated Lck functioning as an adapter protein. Based on the recently available crystal

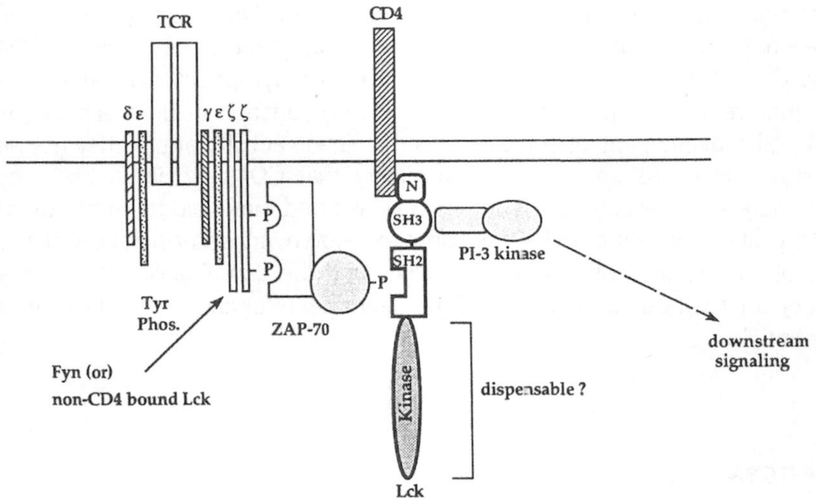


**Fig. 2.** Model for how CD4-associated Lck may function as an adapter molecule. While the interaction of Lck with CD4 is essential for CD4 coreceptor activity, recent data suggest that the kinase activity of CD4-associated Lck may not be necessary. It has also been shown that the SH3 domain of Lck interacts with PI-3 kinase and its SH2 domain interacts with phosphorylated ZAP-70. One possible model for integrating these observations is depicted here. Upon T cell receptor (*TCR*) cross-linking, either Fyn or non-CD4 associated Lck may phosphorylate the TCR components (such as ζ) and ZAP-70 (which interacts with phosphorylated ζ and ε chains). Subsequently CD4-associated Lck via its SH2 domain may interact with ZAP-70. At the same time, Lck may bring in other molecules to the TCR/CD3 complex (such as PI-3 kinase) via its SH3 domain, which could be responsible for the coreceptor activity through CD4

structure of the Lck SH2 and SH3 domains, in which head-to-tail dimers of Lck were observed (Eck et al. 1994), it is tempting to speculate that, if one of the molecules in the Lck dimer binds to ZAP-70, the second molecule may act to bring other proteins into the complex. Such a model would be consistent with the finding that the kinase activity of Lck associated with CD4 is not essential, albeit the association of Lck with CD4 is essential for the coreceptor activity of CD4. However, several aspects of this model are speculative and remain to be tested experimentally.

## 4 Summary

The CD4 molecule plays an important role in the development of CD4⁺T lymphocytes and it also acts as a coreceptor to enhance responses mediated via the TCR. It is now established that CD4 functions both as an adhesion molecule favoring the T cell: APC interaction and as a signaling molecule. The coreceptor function mediated via CD4 depends on its association with Lck, a src-family tyrosine kinase. Lck, while interacting via its unique NH2-terminal domain with CD4, also interacts via its SH2 and SH3 domains with other intracellular signaling proteins. Although the Lck association with CD4 is essential for CD4 coreceptor activity, the tyrosine kinase activity of CD4-associated Lck appears to be dispensable for CD4 function. Given the necessity of Lck kinase activity for T lymphocyte development and for mature T cell functions, perhaps Lck may function at different stages during T cell activation and at some stages the kinase activity of Lck may not be necessary. This raises an intriguing possibility that CD4-associated Lck may function more as an adapter protein than a kinase and may help to recruit other signaling proteins into the TCR/CD3 complex. However, determination of the precise role of Lck in CD4 coreceptor activity and the domains of Lck that are necessary for CD4-dependent and CD4-independent functions awaits further experiments.

## References

Abraham N, Miceli MC, Parnes JR, Veillette A (1991) Enhancement of T cell responsiveness by the lymphocyte-specific tyrosine protein kinase $p56^{lck}$. Nature 350: 62–66

Amrein KE, Flint N, Panholzer B, Burn P (1992) Ras GTPase-activating protein: a substrate and a potential binding protein of the protein-tyrosine kinase $p56^{lck}$. Proc Natl Acad Sci USA 89: 3343–3346

Anderson P, Blue M-L, Morimoto C, Schlossman SF (1987) Cross-linking of T3(CD3) with T4 (CD4) enhances the proliferation of resting T lymophocytes. J Immunol 139: 678–682

Baldari CT, Macchia G, Telford JL (1992) Interleukin-2 promoter activation in T cell expressing activated Ha-ras. J Biol Chem. 267: 4289–4291

Bank I, Chess L (1985) Perturbation of the T4 molecule transmits a negative signal to T cells. J Exp Med 162: 1294–1303

Barber EK, Dasgupta JD, Schlossman SF, Trevillyan JM, Rudd CE (1989) The CD4 and CD8 antigens are coupled to a protein-tyrosine kinase (p56$^{lck}$) that phosphorylates the CD3 complex. Proc Natl Acad Sci USA 86: 3277–3281

Biddison WE, Rao PE, Talle MA, Goldstein G, Shaw S (1984) Possible involvement of the T4 molecule in T cell recognition of class II HLA antigens. Evidence from studies of CTL-target cell binding. J Exp Med 159: 783–797

Buday L, Downward J (1993) Epidermal growth factor regulates p21ras through the formation of a complex of receptor, Grb2 adapter protein and Sos nucleotide exchange factor. Cell 73: 611–620

Buday L, Egan SE, Viciana PR, Cantrell DA, Downward J (1994) A complex of Grb2 adaptor protein Sos exchange factor and a 36 kDa membrane-bound tyrosine phosphoprotein is implicated in Ras activation in T cells. J Biol Chem 269: 9019–9023

Cammarota G, Scheirle A, Takacs B, Doran DM, Knorr R, Bannwarth W, Guardiola J, Sinigaglia F (1992) Identification of a CD4 binding site on the β2 domain of HLA-DR molecules. Nature 356: 799–801

Chen J, Martin BL, Brautigan DL (1992) Regulation of protein serine-threonine phosphatase type 2A by tyrosine phosphorylation. Science 257: 1261–1264

Collins TL, Burakoff SJ (1993) Tyrosine kinase activity of CD4-associated p56$^{lck}$ may not be required for CD4-dependent T-cell activation. Proc Natl Acad Sci USA 90: 11885–11889

Collins TL, Uniyal S, Shin J, Strominger JL, Mittler RS, Burakoff SJ (1992) p56$^{lck}$ association with CD4 is required for the interaction between CD4 and the TCR/CD3 complex and for optimal antigen stimulation. J Immunol 148: 2159–2162

Cooke MP, Abraham KM, Forbush KA, Perlmutter RM (1991) Regulation of T cell receptor signaling by a src family protein-tyrosine kinase (p59fyn) Cell 65: 281–291

Davidson D, Chow LML, Fournel M, Veillette A (1992) Differential regulation of T cell antigen responsiveness by isoforms of the src-related tyrosine protein kinase p59fyn. J Exp Med. 175: 1483–1492

Dialynas DP, Quan ZS, Wall KA, Pierres A, Quintas J, Loken MR, Pierres M, Fitch FW (1993) Characterization of the murine T cell surface molecule, designated L3T4, identified by monoclonal antibody GK1.5: similarity of L3T4 to the human Leu-3/T4 molecule. J Immunol 131: 2445–2551

Dianzani U, Shaw A, Al-Ramadi BK, Kubo RT, Janeway CA (1992) Physical association of CD4 with the T cell receptor. J Immunol 148: 678–688

Downward J (1992) Regulatory mechanisms for ras proteins. Bioessays 14: 177–184

Doyle C, Strominger JL (1987) Interaction between CD4 and class II MHC molecules mediates cell adhesion. Nature 330: 256–259

Duplay P, Thome M, Hervé F, Acuto O (1994) p56$^{lck}$ interacts via its src homology domain with the ZAP-70 kinase. J Exp Med 179: 1163–1172

Eck MJ, Atwell SK, Shoelson SE, Harrison SC (1994) Structure of the regulatory domains of the Src-family tyrosine kinase Lck. Nature 368: 764-769

Egan SE, Giddings BW, Brooks MW, Buday L, Sizeland AM, Weinberg RA (1993) Association of Sos Ras exchange protein with Grb2 is implicated in tyrosine kinase signal transduction and transformation. Nature 363: 45–51

Eichmann K, Jonsson JI, Falk I, Emmrich R (1987) Effective activation of resting mouse T lymphocytes by cross-linking submitogenic concentrations of the T cell antigen receptor with either Lyt-2 or L3T4. Eur J Immunol 17: 643–650

Emmrich F, Strittmatter U, Eichmann K (1986) Synergism in the activation of human CD8 T cells by cross-linking the T-cell receptor complex with the CD8 differentiation antigen. Proc Natl Acad Sci USA 83: 8298–8302

Emmrich F, Kanz L, Eichmann K (1987) Cross-linking of the T cell receptor complex with the subset-specific differentiation antigen stimulates interleukin 2 receptor expression in human CD4 and CD8 T cells. Eur J Immunol 17: 529–534

Gay D, Maddon P, Sekaly R, Talle MA, Godfrey M, Long E. Goldstein G, Chess L, Axel R, Kappler J, Marrack P (1987) Functional interaction between human T-cell protein CD4 and the major histocompatibility complex HLA-DR antigen. Nature 328: 626–629

Gay D, Buus S, Pasternak J, Kappler J, Marrack P (1988) The T-cell accessory molecule CD4 recognizes a monomorphic determinant on isolated Ia. Proc Natl Acad Sci USA 85: 5629–5633

Glaichenhaus N, Shastri N, Littman DR, Turner JM (1991) Requirement for association of p56$^{lck}$ with CD4 in antigen-specific signal transduction in T cells. Cell 64: 511–620

Greenstein JL, Malissen B, Burakoff SJ (1985) Role of L3T4 in antigen-driven activation of a class I-specific T cell hybridoma. J Exp Med. 162: 369–374

Gulbins E, Coggeshall KM, Baier G, Katzav S, Burn P, Altman A (1993) Tyrosine kinase-stimulated guanine nucleotide exchange activity of vav in T cell activation. Science 260: 822–825
Hsi ED, Isegel JN, Minami Y, Luong ET, Klausner RD, Samelson LE (1989) T cell activation induces rapid tyrosine phosphorylation of a limited number of cellular substrates. J Biol Chem 264: 10836–10842
Iwashima M, Irving BA, Vanoers NSC, Chan AC, Weiss A (1994) Sequential interactions of the TCR with two distinct cytoplasmic tyrosine kinases. Science 263: 1136–1139
Izquierdo M, Downward J, Graves JD, Cantrell DA (1992) Role of protein kinase C in T-cell antigen receptor regulation of $p21^{ras}$ evidence that two $p21^{ras}$ regulatory pathways coexist in T cells. Mol Cell Biol 12: 3305–3312
June CH, Fletcher MC, Ledbetter JA, Schieven GL, Siegel JN, Phillips AF, Samelson LE (1990a) Increases in tyrosine phosphorylation are detectable before phospholipase C activation after T cell receptor stimulation. J Immunol 144: 1591–1599
June CH, Fletcher MC, Ledbetter JA, Schieven GL, Siegel JN, Phillips AF, Samelson LE (1990b) Inhibition of tyrosine phosphorylation prevents T-cell receptor-mediated signal transduction. Proc Natl Acad Sci USA 87: 7722–7726
Kapeller R, Prasad KVS, Janssen O, Hou W, Schaffhausen BS, Rudd CE, Cantley LC (1994) Identification of two SH3-binding motifs in the regulatory subunit of phosphatidylinositol 3-kinase. J Biol Chem 269: 1927–1933
Karnitz L, Sutor SL, Torigoe T, Reed JC, Bell MP, McKean DJ, Leibson PJ, Abraham RT (1992) Effects of $p56^{lck}$ deficiency on the growth and cytolytic effector function of an interleukin-2-dependent cytotoxic T-cell line. Mol Cell Biol 12: 4521–4530
Killeen N, Littman DR (1993) Helper T-cell development in the absence of CD4-$p56^{lck}$ association. Nature 364: 729–732
König R, Huang LY, Germain RN (1992) MHC class II interaction with CD4 mediated by a region analogous to the MHC class I binding site for CD8. Nature 356: 796–799
Krensky AM, Clayberger C, Reiss CS, Strominger JL, Burakoff SJ (1982) Specificity of OKT4+ cytotoxic T lymphocyte clones. J Immunol 129: 2001–2003
Kupfer A, Singer SJ, Janeway CA, Swain SL (1987) Coclustering of CD4 (L3T4) molecule with the T-cell receptor is induced by specific direct interaction of helper T cells and antigen-presenting cells. Proc Natl Acad Sci USA 84: 5888–5892
Lorenz U, Ravichandran KS, Pei D, Walsh CT, Burakoff SJ, Neel BG (1994) Lck-dependent tyrosyl phosphorylation of the phosphotyrosine phosphatase SH-PTP1 in murine T cells. Mol Cell Biol 14: 1824–1834
Lowenstein EJ, Daly RJ, Batzer AG, Li W, Margolis B, Lammers R, Ullrich A, Skolnik EY, Bar-Sagi D, Schlessinger J (1992) The SH2 and SH3 domain-containing protein GRB2 links receptor tyrosine kinases to ras signaling. Cell 70: 431–442
Luo K, Sefton BM (1990) Cross-linking of T-cell surface molecules CD4 and CD8 stimulates phosphorylation of the lck tyrosine protein kinase at the autophosphorylation site. Mol Cell Biol 10: 5305–5313
Marrack P, Enders R, Shimonkevitz R, Zlotnik A, Dialynas D, Fitch F, Kappler J (1983) The major histocompatibility complex-restricted antigen receptor on T cells. II. Role of the L3T4 product. J Exp Med 158: 1077–1091
Marth JD, Lewis DB, Cooke MP, Mellins ED, Gearn ME, Samelson LE, Wilson CB, Miller AD, Perlmutter RM (1989) Lymphocyte activation provokes modification of a lymphocyte-specific protein tyrosine kinase ($p56^{lck}$) J Immunol 142: 2430–2437
McGlade J, Cheng A, Pelicci G, Pelicci PG, Pawson T (1992) Shc proteins are phosphorylated and regulated by the v-Src and v-Fps protein-tyrosine kinases. Proc Natl Acad Sci USA 89: 8869–8873
Mittler RS, Goldman SJ, Spitalny GL, Burakoff SJ (1989) T-cell receptor-CD4 physical association in a murine T-cell hybridoma: induction by antigen receptor ligation. Proc Natl Acad Sci USA 86: 8531–8535
Molina TJ, Kishihara K, Siderovski DP, van Ewijk W, Narendran A, Timms E, Wakeham A, Paige CJ, Hartmann K-U, Veillette A, Davidson D, Mak TW (1992) Profound block in thymocyte development in mice lacking p56lck. Nature 357: 161–164
O'Shea JJ, Ashwell JD, Bailey TL, Cross SL, Samelson LE, Klausner RD (1991) Expression of v-src in a murine T-cell hybridoma results in constitutive T-cell receptor phosphorylation and interleukin 2 production. Proc Natl Acad Sci USA 88: 1741–1745
Owens T, Fazekas de St Groth B, Miller JFAP (1987) Coaggregation of the T-cell receptor with CD4 and other T-cell surface molecules enhances T-cell activation. Proc Natl Acad Sci USA 84: 9209–9213

Park DJ, Rho HW, Rhee SG (1991) CD3 stimulation causes phosphorylation of phospholipase C-gamma 1 on serine and tyrosine residues in a human T-cell line. Proc Natl Acad Sci USA 88: 5453–5456

Pawson T, Gish GD (1992) SH2 and SH3 domains: from structure to function. Cell 71: 359–362

Pelicci G, Lanfrancone L, Grignani F, McGlade J, Cavallo F, Forni G, Nicoletti I, Grignani F, Pawson T, Pelicci PG (1992) A novel transforming protein (SHC) with an SH2 domain is implicated in mitogenic signal transduction. Cell 70: 93–104

Portoles P, Rojo J, Golby A, Bonneville M, Gromkowski S, Greenbaum L, Janeway CA, Murphy DB, Bottomly K (1989) Monoclonal antibodies to murine CD3ε define distinct epitopes, one of which may interact with CD4 during T cell activation. J Immunol 142: 4169–4175

Prasad KVS, Janssen O, Kapeller R, Raab M, Cantley LC, Rudd CE (1993a) Src-homology 3 domain of protein kinase p59$^{fyn}$ mediates binding to phosphatidylinositol 3-kinase in T cells. Proc Natl Acad Sci USA 90: 7366–7370

Prasad KVS, Kapeller R, Janssen O, Rapke H, Duke-Cohan JS, Cantley LC, Rudd CE (1993b) Phosphatidylinositol (PI) 3-kinase and PI 4-kinase binding to the CD4-p56lck complex: the p56lck SH3 domain binds to PI 3-kinase but not PI 4-kinase. Mol Cell Biol 13: 7708–7717

Ravichandran KS, Lee KK, Songyang Z, Cantley LC, Burn P, Burakoff SJ (1993) Interaction of Shc with ζ chain of the T cell receptor upon T cell activation. Science 262: 902–905

Ravichandran KS, Lorenz V, Shoelson SE, Burakoff SJ (1995) Interaction of Shc with Grb2 regulates the association of Grb2 with mSOS. Mol Cell Biol 15: 593–600

Rayter S, Woodrow M, Lucas S, Cantrell D, Downward J (1992) p21ras mediates control of IL2 gene promoter function in T cell activation. EMBO J 11: 4549–4556

Reinherz EL, Kung PC, Goldstein G, Schlossman SF (1979) Separation of functional subsets of human T cells by a monoclonal antibody. Proc Natl Acad Sci USA 76: 4061–4065

Ren R, Mayer BJ, Cicchetti P, Baltimore D (1993) Identification of a ten-amino acid proline-rich SH3 binding site. Science 259: 1157–1161

Rogozinski L, Bass A, Glickman E, Talle MA, Goldstein G, Wang J, Chess L, Thomas Y (1984) The T4 surface antigen is involved in the induction of helper function. J Immunol 132: 735–739

Rojo JM, Saizawa K, Janeway CA (1989) Physical association of CD4 and the T-cell receptor can be induced by anti-T-cell receptor antibodies. Proc Natl Acad Sci USA 86: 3311–3315

Rozakis-Adcock M, McGlade J, Mbamalu G, Pelicci G, Daly R, Li W, Batzer A, Thomas S, Brugge J, Pelicci PG (1992) Association of the Shc and Grb2/Sem5 SH2-containing proteins is implicated in activation of the Ras pathway by tyrosine kinases. Nature 360: 689–692

Rudd CE, Trevillyan JM, Dasgupta JD, Wong LL, Schlossman SF (1988) The CD4 receptor is complexed in detergent lysates to a protein-tyrosine kinase (pp58) from human T lymphocytes. Proc Natl Acad Sci USA 85: 5190–5194

Saizawa K, Rojo J, Janeway CA (1987) Evidence for a physical association of CD4 and the CD3: α: β T-cell receptor. Nature 328: 260–263

Samelson LE, Phillips AF, Luong ET, Klausner RD (1990) Association of the fyn protein-tyrosine kinase with the T-cell antigen receptor. Proc Natl Acad Sci USA 87: 4358–4362

Schlessinger J (1993) How receptor tyrosine kinase activate Ras.Trends Biol Sci 18: 273–275

Shaw AS, Amrein KE, Hammond C, Stern DF, Sefton BM, Rose JK (1989) The lck tyrosine protein kinase interacts with the cytoplasmic tail of the CD4 glycoprotein through its unique amino-terminal domain. Cell 59: 627–636

Shaw AS, Chalupny J, Whitney JA, Hammond C, Amrein KE, Kavathas P, Sefton BM, Rose JK (1990) Short related sequences in the cytoplasmic domains of CD4 and CD8 mediate binding to the amino-terminal domain of the p56$^{lck}$ tyrosine protein kinase. Mol Cell Biol 10: 1853–1862

Shultz LD, Schweitzer PA, Rajan TV, Yi T, Ihle JN, Matthews RJ, Thomas ML, Beier DR (1993) Mutations at the murine motheaten locus are within the hematopoietic cell protein phosphatase (HCHP) gene. Cell 73: 1445–1454

Sieh M, Batzer A, Schlessinger J, Weiss A (1994) GRB2 and Phospholipase C-g1 associate with a 36- to 38-Kilodalton phosphotyrosine protein after T-cell receptor stimulation. Mol Cell Biol 14: 4435–4442

Sleckman BP, Peterson A, Jones WK, Foran JA, Greenstein JL, Seed B, Burakoff SJ (1987) Expression and function of CD4 in a murine T-cell hybridoma. Nature 328: 351–353

Sleckman BP, Peterson A, Foran JA, Gorga JC, Kara CJ, Strominger JL, Burakoff SJ, Greenstein JL (1988) Functional analysis of a cytoplasmic domain-deleted mutant of the CD4 molecule. J Immunol 141: 49–54

Sleckman BP, Rosenstein Y, Igras VE, Greenstein JL, Burakoff SJ (1991) Glycolipid-anchored form of CD4 increases intercellular adhesion but is unable to enhance T cell activation. J Immunol 147: 428–431

Songyang Z, Shoelson SE, Chaudhuri M, Gish G, Pawson T, Haser WG, King F, Roberts T, Ratnofsky S, Lechleider RJ, Neel BG, Birge RB, Fajardo JE, Chou MM, Hanafusa H, Schaffhausen B, Cantley LC (1993) SH2 domains recognize specific phosphopeptide sequences. Cell 72: 767–778

Spits H, Borst J, Terhorst C, de Vries JE (1982) The role of T cell differentiation markers in antigen-specific and lectin-dependent cellular cytotoxicity mediated by T8+ and T4+ human cytotoxic T cell clones directed at class I and class II MHC antigens. J Immunol 129: 1563–1569

Stein PH, Fraser JD, Weiss A (1994) The cytoplasmic domain of CD28 is both necessary and sufficient for costimulation of interleukin-2 secretion and association with phosphatidylinositol 3'-kinase. Mol Cell Biol 14: 3392–3402

Straus DB, Weiss A (1992) Genetic evidence for the involvement of the lck tyrosine kinase in signal transduction through the T cell antigen receptor. Cell 70: 585–593

Terhorst C, van Agthoven A, Reinherz E, Schlossman SF (1980) Differentiation antigens T4 and T5. Science 209: 520–521

Truitt KE, Hicks CM, Imboden JB (1994) Stimulation of CD28 triggers an association between CD28 and phosphatidylinositol-3-kinase in Jurkat T cells. J Exp Med 179: 1071–1076

Turner JM, Brodsky MH, Irving BA, Levin SD, Perlmutter RM, Littman DR (1990) Interaction of the unique N-terminal region of tyrosine kinase $p56^{lck}$ with cytoplasmic domains of CD4 and CD8 is mediated by cysteine motifs. Cell 60: 755–765

Valius M, Kazlauskas A (1993) Phospholipase C-γ1 and phosphatidylinositol 3 kinase are the downstream mediators of the PDGF receptor's mitogenic signal. Cell 73: 321–334

Vega MA, Kuo M-C, Carrera AC, Strominger JL (1990) Structural nature of the interaction between T lymphocyte surface molecule CD4 and the intracellular protein tyrosine kinase lck. Eur J Immunol 20: 453–456

Veillette A, Bookman MA, Horak EM, Bolen JB (1988) The CD4 and CD8 T cell surface antigens are associated with the internal membrane tyrosine-protein kinase $p56^{lck}$. Cell 55: 301–308

Veillette A, Bolen JB, Bookman MA (1989a) Alterations in tyrosine protein phosphorylation induced by antibody-mediated cross-linking of the CD4 receptor of T lymphocytes. Mol Cell Biol 9: 4441–4446

Veillette A, Bookman MA, Horak EM, Samelson LE, Bolen JB (1989b) Signal transduction through the CD4 receptor involves the activation of the internal membrane tyrosine-protein kinase $p56^{lck}$. Nature 338: 257–259

Veillette A, Sleckman BP, Ratnofsky S, Bolen JB, Burakoff SJ (1990) The cytoplasmic domain of CD4 is required for stable association with the lymphocyte-specific tyrosine kinase $p56^{lck}$. Eur J Immunol 20: 1397–1400

Walker C, Bettens F, Pickler WJ, (1987) Activation of T cells by cross-linking an anti-CD3 antibody with a second anti-T cell antibody: mechanism and subset-specific activation. Eur J Immunol 17: 873–880

Wange R, Malek S, Desiderio S, Samelson L (1993) Tandem SH2 domains of ZAP-70 bind to T cell antigen receptor ζ and CD3ε from activated Jurkat T cells. J Biol Chem 268: 19797–19801

Wassmer P, Chan C, Logdberg L, Shevach EM (1985) Role of the L3T4-antigen in T cell activation. II. Inhibition of T cell activation by monoclonal anti-L3T4 antibodies in the absence of accessory cells. J Immunol 135: 2237–2242

Webb M, Mason DW, Williams AF (1979) Inhibition of mixed lymphocyte response by monoclonal antibody specific for a rat T lymphocyte subset. Nature 282: 841–843

Weber JR, Bell GM, Han MY, Pawson T, Imboden JB (1992) Association of the tyrosine kinase Lck with phospholipase Cγ1 after stimulation of the T cell antigen receptor. J Exp Med 176: 373–379

Weiss A, Koretzky G, Schatzman R, Kadlecek T (1991) Stimulation of the T cell antigen receptor induces tyrosine phosphorylation of phospholipase Cγ1. Proc Natl Acad Sci USA 88: 5484–5488

Wilde DB, Marrack P, Kappler J, Dialynas DP, Fitch FW (1983) Evidence implicating L3T4 in class II MHC antigen reactivity; monoclonal antibody GK15 (anti-L3T4a) blocks class II MHC antigen-specific proliferation, release of lymphokines, and binding by cloned murine helper T lymphocyte lines. J Immunol 131: 2178–2183

Xu H, Littman DR (1993) A kinase-independent function of Lck in potentiating antigen-specific T cell activation. Cell 74: 633–643

# Signal Transduction by the Lymphocyte-Specific Tyrosine Protein Kinase p56$^{lck}$

R. WEIL[1] and A. VEILLETTE[1,2,3]



## 1 Introduction

Stimulation of T Lymphocytes by antigen or anti-T cell receptor (TCR) antibodies provokes a rapid series of intracellular biochemical changes, leading to the expression of either the helper-inducer or the cytotoxic T cell phenotype (for reviews, see SAMELSON and KLAUSNER 1992; VEILLETTE and DAVIDSON 1992; PERLMUTTER et al. 1993 and WEISS and LITTMAN 1994). This response, termed T cell

[1]McGill Cancer Centre, McIntyre Medical Sciences Building, McGill University, 3655 Drummond Street, Montreal, Canada H3G IY6
[2]Departments of Biochemistry, Medicine, and Oncology, McIntyre Medical Sciences Building, McGill University, 3655 Drummond Street, Montreal, Canada H3G IY6
[3]Departments of Medicine and Oncology, Montreal General Hospital, Montreal, Canada H3G 1A4

activation, is accompanied by a sudden increase in intracellular phosphotyrosine levels. Importantly, the use of tyrosine protein kinase inhibitors has revealed that tyrosine protein phosphorylation is required for all subsequent events of T cell activation, including stimulation of phosphatidyl inositol (PI) turnover, rise in intracellular calcium, lymphokine secretion and mitogenesis (MUSTELIN et al. 1990; JUNE et al. 1990).

As reviewed elsewhere (SAMELSON and KLAUSNER 1992; WEISS and LITTMAN 1994), the TCR and the associated CD3 and zeta (ζ) chains, do not possess intrinsic catalytic properties capable of causing tyrosine protein phosphorylation. Nevertheless, the CD3 and ζ molecules bear, within their intracellular domains, multiple copies of an immunoreceptor tyrosine-based activation motif (ITAM; RETH 1989). This motif functions by orchestrating interactions between cytoplasmic tyrosine protein kinases and tyrosine phosphorylation substrates. While the exact mechanism by which ITAMS regulate tyrosine protein phosphorylation is not known, recent studies have identified the kinases recruited in this process (for reviews, see SAMELSON and KLAUSNER 1992; VEILLETTE and DAVIDSON 1992; PERLMUTTER et al. 1993 and WEISS and LITTMAN 1994). Two members of the Src family of tyrosine protein kinases abundantly expressed in T cells, $p56^{lck}$ and $p59^{fynT}$, have been firmly implicated in the *initiation* of antigen receptor-induced signal transduction. Moreover, Zap-70, a tyrosine protein kinase which forms a complex with the antigen receptor during T cell activation, apparently plays a role in the *amplification* of the TCR-induced tyrosine protein phosphorylation signal.

This chapter reviews the regulation and function of the lymphocyte-specific tyrosine protein kinase $p56^{lck}$. Considerable interest in this enzyme stems in part from its physical association with the CD4 and CD8 T cell surface antigens. Moreover, recent data show that $p56^{lck}$ also plays a central role in antigen receptor-mediated signal transduction. (For detailed discussions of the regulation and function of FynT and Zap-70, interested readers are referred to SAMELSON and KLAUSNER 1992; VEILLETTE and DAVIDSON 1992; PERLMUTTER et al. 1993 and WEISS and LITTMAN 1994)

## 2 $p56^{lck}$ Is a Lymphocyte-Specific Src-Related Tyrosine Protein Kinase

T lymphocytes express several distinct tyrosine protein kinases. These include classical receptor tyrosine protein kinases (such as the insulin receptor), which are activated by extracellular ligands. Unfortunately, little is known about the functions of these enzymes in normal T cell physiology. T cells also contain non-receptor tyrosine protein kinases such as Lck, FynT, Zap-70 and Csk. Whereas these enzymes do not possess an external domain for direct interactions with the extracellular milieu, several are capable of direct or indirect associations with the inner aspect of the plasma membrane. This property is presumably crucial for the recognized ability of these products to participate in the signal transduction of T cell activation.

The Src family of tyrosine protein kinases comprises nine distinct members: c-Src, Fyn, c-Yes, Yrk, Lck, Lyn, c-Fgr, Hck and Blk (for a review, see COOPER 1990). As the genes coding for these products possess similar exon-intron structures, they are likely derived from a common ancestor. Like other Src-related polypeptides, $p56^{lck}$ bears multiple independent structural domains (Fig. 1). These include, from the $NH_2$-terminus to the COOH-terminus: (a) $NH_2$-terminal sites of myristylation (glycine 2) and palmitylation (cysteines 3 and/or 5) (SHENOY-SCARIA et al. 1993), which allow stable association with the inner aspect of the plasma membrane; (b) a "unique" domain of roughly 60 amino acids, conferring to $p56^{lck}$ its exquisite ability to associate with the CD4 and CD8 T cell surface molecules (VEILLETTE et al. 1988b; RUDD et al. 1988; SHAW et al. 1989, 1990; TURNER et al. 1990); (c) a Src homology (SH) 3 motif, permitting associations with proline-rich proteins involved in signal amplification (reviewed in PAWSON and GISH 1992); (d) an SH2 domain, having the demonstrated ability to bind phosphotyrosine-containing peptides (reviewed in KOCH et al. 1991; CANTLEY et al. 1991 and PAWSON and GISH 1992); (e) a catalytic domain, including sites for ATP binding, phosphotransfer (lysine 273) and autophosphorylation (tyrosine 394) and (f) a COOH-terminal negative regulatory domain, bearing the major site of in vivo tyrosine phosphorylation (tyrosine 505).

Within normal cell populations, $p56^{lck}$ is exclusively expressed in the lymphoid lineage, especially in T lymphocytes (MARTH et al. 1985; VEILLETTE et al. 1987). Analyses of purified T cell subsets have revealed that $p56^{lck}$ abounds in all immature and mature T cell types (VEILLETTE et al. 1989c; REYNOLDS et al. 1990). High amounts of Lck protein are also observed in natural killer (NK) cells (EINSPHAR et al. 1990). Moreover, while low or undetectable quantities of $p56^{lck}$ are present

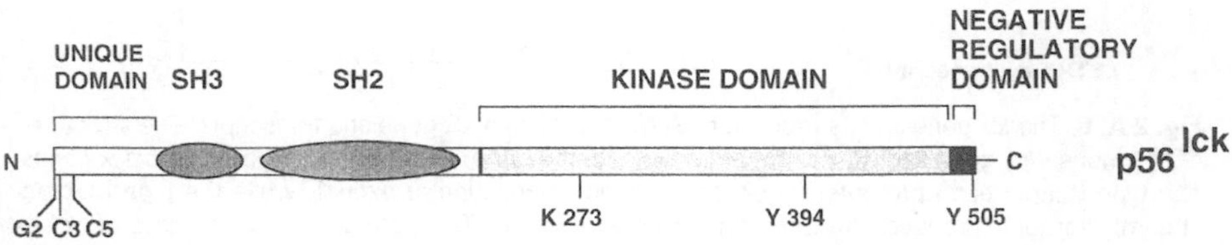


**Fig. 1.** Primary structure of $p56^{lck}$. Glycine 2 (*G2*) is the $NH_2$-terminal site of Lck myristylation, while cysteines 3 (*C3*) and/or 5 (*C5*) are its sites of palmitylation. The unique domain allows specific binding of $p56^{lck}$ to the CD4 and CD8 T-cell surface antigens (see **Fig. 4.**) The SH3 and SH2 regions are conserved noncatalytic sequences of approximately 50 and 100 amino acids in length, respectively. While the SH3 region mediates interactions with proline-rich polypeptides, the SH2 motif allows associations with phosphotyrosine-containing proteins. The catalytic domain of $p56^{lck}$ contains a phosphotransfer motif, centered around lysine 273 (*K273*), and the major site of autophosphorylation, tyrosine 394 (*Y394*). The COOH-terminal negative regulatory sequence bears the major site of in vivo tyrosine phosphorylation of $p56^{lck}$, tyrosine 505

in resting B cells, significant expression is noted in activated B cells (TAIEB et al. 1993).

The lymphocyte-specific expression of p56[lck] is conferred by the concerted actions of two different promoters, separated by approximately 35 kilobases (kb) in the human genome (WILDIN et al. 1991) (Fig. 2A). The upstream (type II) *lck* promoter is active in both immature and mature T lymphocytes (REYNOLDS et al. 1990; WILDIN et al. 1991). In contrast, the downstream (type I) promoter is mostly functional in immature T cells. These two promoters dictate the production of two distinct transcripts which diverge only within their 5' untranslated sequences. Interestingly, even though these transcripts code for identical Lck

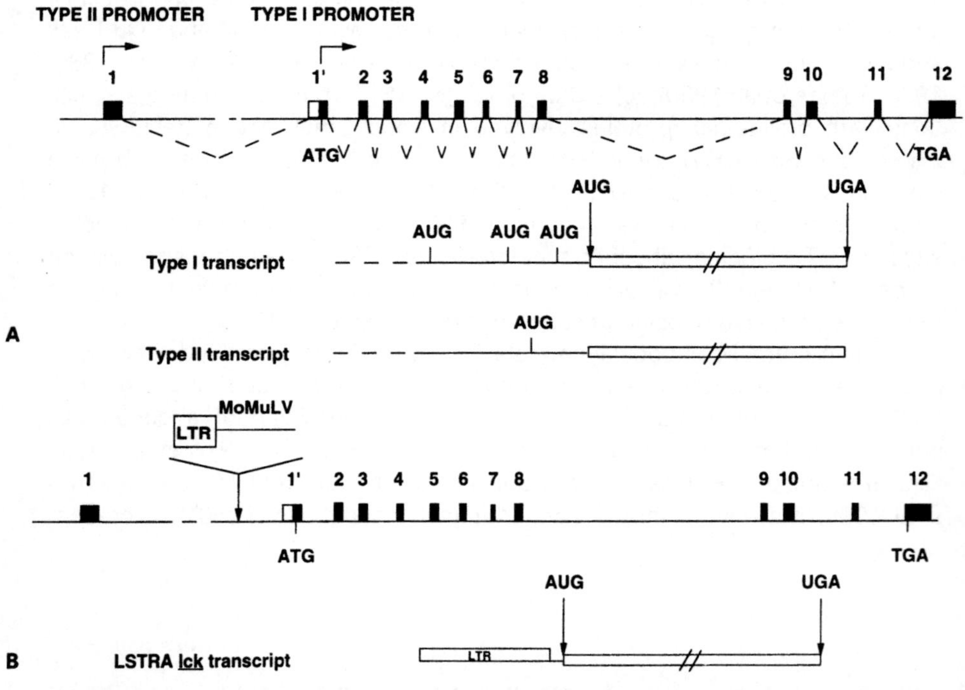


**Fig. 2 A, B.** The *lck* gene and its transcripts. **A** Normal murine *lck* gene and transcripts. The structures of the mouse *lck* gene and transcripts are shown. In the gene, exons are highlighted as *black boxes.* The type II (upstream) promoter regulates transcription initation at exon 1, while the type I (downstream) promoter produces transcripts initiated at exon 1'. The pattern of RNA splicing of *lck* is indicated by a *dashed line.* Note that, when exon 1 is used, it is spliced to an acceptor site located in the body of exon 1'. The structures of the two types of *lck* transcripts are indicated *below.* The locations of the initiator (ATG/AUG) and stop (TGA/UGA) codons are shown. Three AUGs are present in the 5' UT region of the type I transcript, while one is located in the equivalent region of the type II RNA. The upstream AUGs may be involved in translational regulation of *lck.* **B** LSTRA *lck* gene and transcript. The structures of the rearranged *lck* allele in LSTRA and of the corresponding transcript are depicted. As a consequence of insertion of a Moloney murine leukemia virus (*MoMuLV*) long terminal repeat (*LTR*) in the 5' portion of the *lck* gene, a fusion transcript is produced in which the 5' UT of the *lck* RNA in replaced by viral LTR sequences. This gene rearrangement causes a sevenfold increase in *lck* transcription and a seven- to tenfold elevation of *lck* mRNA translation. Combined, these changes lead to a 50-fold augmentation of the abundance of p56[lck] in LSTRA cells

polypeptides, their different 5' ends have been postulated to confer upon them distinct susceptibilities to translational regulation (MARTH et al. 1988b).

# 3 Regulation of p56$^{lck}$

## 3.1 Negative Regulation by Phosphorylation of Tyrosine 505

As is the case for other Src-related enzymes, the function of p56$^{lck}$ is primarily regulated by tyrosine phosphorylation. Its catalytic activity is repressed by phosphorylation of a conserved COOH-terminal tyrosine residue, tyrosine 505 (MARTH et al. 1988a; AMREIN and SEFTON 1988; ABRAHAM and VEILLETTE 1990). Presumably, this inhibitory effect results from the ability of phosphorylated tyrosine 505 to participate in an intramolecular association with the SH2 domain, thereby precluding interactions of the catalytic region with intracellular substrates (Fig. 3) (reviewed in CANTLEY et al. 1991; KOCH et al. 1991 and PAWSON and GISH 1992). This model is supported by the finding that synthetic peptides corresponding to the phosphorylated COOH-terminal of p56$^{lck}$ can interact with recombinant Lck SH2 domains in vitro (AMREIN et al. 1993; PAYNE et al. 1993; PERI et al. 1993). Furthermore, it is consistent with the observations that mutation of tyrosine 505 to a non-phosphorylatable phenylalanine, or deletion of the SH2 domain, activates the catalytic function of Lck in vivo, thereby uncovering its oncogenic potential towards rodent fibroblasts (AMREIN and SEFTON 1988; MARTH et al. 1988a; ABRAHAM

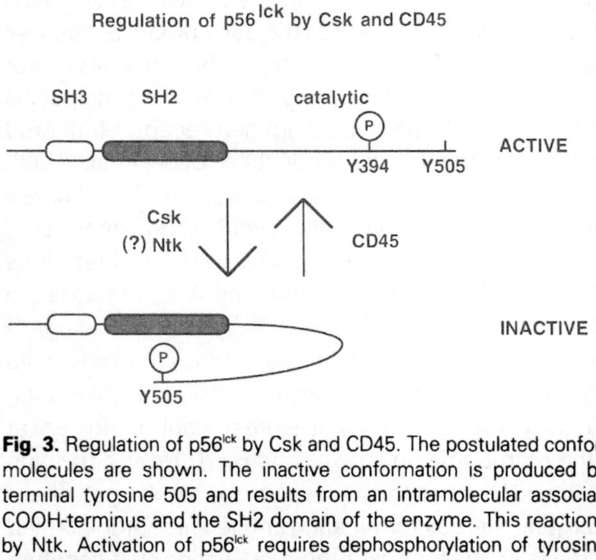


**Fig. 3.** Regulation of p56$^{lck}$ by Csk and CD45. The postulated conformations of active and inactive Lck molecules are shown. The inactive conformation is produced by phosphorylation of the COOH-terminal tyrosine 505 and results from an intramolecular association between the phosphorylated COOH-terminus and the SH2 domain of the enzyme. This reaction is catalyzed by Csk and, perhaps, by Ntk. Activation of p56$^{lck}$ requires dephosphorylation of tyrosine 505, which is mediated by the transmembrane tyrosine protein phosphatase CD45. Upon dephosphorylation, p56$^{lck}$ undergoes autophosphorylation at tyrosine 394, the site of positive regulation

and VEILLETTE 1990; VEILLETTE et al. 1992; REYNOLDS et al. 1992). In addition to regulating $p56^{lck}$ by modifying its conformation, the SH2 sequence functions by stabilizing phosphorylation at tyrosine 505, thus securing its inhibitory impact on the $p56^{lck}$-associated kinase activity (GERVAIS et al. 1993). This supplementary effect correlates with the potential of the SH2 domain to protect phosphorylated tyrosine 505 from tyrosine protein phosphatases.

### 3.1.1 $p50^{csk}$: A Negative Regulator of $p56^{lck}$

The catalytic activity of $p56^{lck}$ is dictated by the balance between the actions of the kinases and phosphatases that regulate tyrosine 505. Phosphorylation at tyrosine 505 is not the result of autophosphorylation, but instead is mediated by another cytoplasmic tyrosine protein kinase, $p50^{csk}$ (Fig. 3) (NADA et al. 1991; BERGMAN et al. 1992; GERVAIS et al. 1993). Like Src family members, Csk possesses NH2-terminal SH3 and SH2 domains and a COOH-terminal catalytic domain (NADA et al. 1991; PARTANEN et al. 1991; SABE et al. 1992). However, in contrast to Src-related polypeptides, it does not bear sites of myristylation and autophosphorylation and does not contain a COOH-terminal site of tyrosine phosphorylation. In fact, because the exon-intron structure of the *csk* gene is unique (BRÄUNINGER et al. 1993), it is implied that Csk belongs to a separate tyrosine protein kinase family.

In vitro studies have demonstrated that $p50^{csk}$ can phosphorylate the COOH-terminal tyrosine of at least five Src-related enzymes: c-Src, Lck, Fyn, Lyn and c-Fgr (NADA et al. 1991; OKADA et al. 1991; BERGMAN et al. 1992; GERVAIS et al. 1993; RUZZENE et al. 1994). On this basis, it is likely that Csk can negatively regulate all members of the Src family. Even though it is ubiquitously expressed, $p50^{csk}$ is most abundant in hematopoietic cells, including T cells (PARTANEN et al. 1991; OKADA et al. 1991; SABE et al. 1992). This finding led to the early proposal that Csk may play an important role in normal T cell physiology. This idea was later supported by the observation that overexpression of $p50^{csk}$ in an antigen-specific mouse T cell hybridoma (BI-141) strongly decreased antigen receptor-triggered tyrosine protein phosphorylation and lymphokine production (CHOW et al. 1993). Importantly, this effect was rescued by expression of Lck or FynT molecules bearing a tyrosine-to-phenylalanine substitution at the negative regulatory site, indicating that Csk does function in vivo by phosphorylating the COOH-terminus of Src-related enzymes (CHOW et al. 1993; L.M.L. Chow, and A.V., unpublished data).

Whereas the mode of regulation of $p50^{csk}$ in T lymphocytes is not known, its function is postulated to be controlled by changes in cellular localization. In resting T cells, Csk is mostly found in the cytoplasm, where it is presumably inefficient at repressing the functions of the membrane-bound Lck and FynT. Upon activation of $p56^{lck}$ and/or $p59^{fynT}$, $p50^{csk}$ would be recruited to the plasma membrane, where it could efficiently act on its targets, thereby terminating antigen receptor-induced signal transduction. This model is consistent with the observation that constitutive targetting of Csk to the plasma membrane by the addition of a Src myristylation signal, or a Ras farnesylation/palmitylation signal, greatly increases its ability to inhibit TCR-mediated signals (CHOW et al. 1993). In a more physiological context, the recruitment of Csk to the membrane could be mediated by its SH3 and SH2 domains, two motifs able to link other signal transduction molecules to the plasma membrane (reviewed in KOCH et al. 1991; CANTLEY et al. 1991 and PAWSON and GISH 1992). In addition to possibly being regulated by reversible membrane association, there is evidence that the function of $p50^{csk}$ may be controlled by changes in its catalytic activity. Indeed, OETKEN et al. (1994) showed that the ability of Csk to phosphorylate an exogenous substrate in vitro is moderately increased after activation of T cells wtih anti-CD3 antibodies. The exact mechanism for this activation remains to be established.

Recently, cDNAs corresponding to a second member of the *csk* family were independently cloned by several groups (BENNETT et al, 1994; SAKANO et al. 1994; KLAGES et al. 1994; CHOW et al. 1994; MCVICAR et al. 1994). Like Csk, the novel enzyme (variably termed Matk, Hyl, Ctk, Ntk and Lsk) can phosphorylate the COOH-terminal tyrosine residue of $p56^{lck}$ in vitro (KLAGES et al. 1994; CHOW et al. 1994). However, contrary to Csk, the Csk-related polypeptide (hereafter termed Ntk) is exclusively expressed in brain and hematopoietic cells, including T cells (BENNETT et al. 1994; SAKANO et al. 1994; KLAGES et al. 1994; CHOW et al. 1994; MCVICAR et al. 1994). In one report, the expression of *ntk* was dramatically reduced following activation of mouse T cells with a variety of stimulatory regimens (CHOW et al. 1994). Paradoxically, however, in another study, stimulation of human peripheral blood lymphocytes with phytohemagglutinin was shown to induce the expression of *ntk* (MCVICAR et al. 1994). While the basis for these differences is not known, these results suggest that Ntk may play a role in the regulation of T cell activation. Future studies will be necessary to confirm this possibility.

### 3.1.2 CD45: A Positive Regulator of $p56^{lck}$

Dephosphorylation of tyrosine 505 is mediated by CD45, a transmembrane tyrosine protein phosphatase abundantly expressed in all nucleated hematopoietic cells (Fig. 3). This conclusion is primarily based on analyses of CD45-deficient T cell variants. Indeed, Lck molecules obtained from CD45-negative T cells show markedly increased (eight- to tenfold) tyrosine 505 phosphorylation, when compared with $p56^{lck}$ polypeptides isolated from their CD45-positive counterparts (OSTERGAARD et al. 1989; HURLEY et al. 1993; CAHIR MCFARLAND et al. 1993; SIEH et al. 1993). CD45-mediated dephosphorylation of tyrosine 505 is likely critical for the initiation of TCR-mediated signal transduction, as indicated by the observation that CD45-deficient T cells fail to be activated by antigen receptor stimulation (PINGEL and THOMAS 1989; KORETZKY et al. 1990; WEAVER et al. 1991; VOLAREVIC et al. 1992). Interestingly, however, this deficiency can be rescued by coligation of TCR with CD4 or CD8, which are known to associate with $p56^{lck}$ (see below; SHIROO et al. 1992; DEANS et al. 1992). Seemingly, engagement of CD4 or CD8 allows a CD45-independent activation of $p56^{lck}$.

In contrast to $p56^{lck}$, FynT molecules isolated from CD45-negative T cells demonstrate a more limited (two- to fourfold) increase in COOH-terminal tyrosine phosphorylation (Hurley et al. 1993; Cahir McFarland et al. 1993). This observation implies that FynT is less efficiently regulated by CD45. This difference may relate to the recent determination that, in T cells, $p59^{fynT}$ is extensively associated with the centrosome, microtubules and mitotic spindle, where it may not be amenable to regulation by CD45 (Ley et al. 1994). It is also possible that, at the membrane, CD45 has a greater affinity for $p56^{lck}$ than for $p59^{fynT}$. This possibility may explain the finding that small quantities of Lck, but not FynT, are associated with CD45 in mild detergent lysates of T lymphocytes (Schraven et al. 1991; Koretzky et al. 1993).

The processes controlling the activity of CD45 in T cells are not determined. Since CD45 carries a long extracellular domain with variable structures as a result of alternative splicing of the CD45 gene, it is plausible that its phosphatase activity is modulated by one or more yet unknown ligands (for a review, see Trowbridge and Thomas 1994). If factual, this level of regulation would add further complexity to the control of $p56^{lck}$-mediated functions in T lymphocytes. Additional studies will be necessary to examine this possibility.

## 3.2 Positive Regulation by Phosphorylation of Tyrosine 394

Tyrosine 394 is the major site of autophosphorylation of $p56^{lck}$. Consequently, this residue undergoes prominent phosphorylation during in vitro immune complex kinase reactions. Tyrosine 394 is not significantly phosphorylated in resting T cells or in NIH 3T3 fibroblasts transfected with wild-type *lck* cDNAs (Veillette et al. 1989a; Amrein and Sefton 1988; Marth et al. 1988a; Abraham and Veillette 1990). Nonetheless, it becomes occupied upon activation of $p56^{lck}$ by mutation of tyrosine 505 or antibody-mediated aggregation of CD4 (Amrein and Sefton 1988; Marth et al. 1988a; Abraham and Veillette 1990; Veillette et al. 1989a; Luo and Sefton 1990a). Importantly, mutation of tyrosine 394 to phenylalanine dramatically reduces the ability of activated $p56^{lck}$ to cause oncogenic transformation of rodent fibroblasts and to enhance T cell responsiveness to antigenic stimulation (Abraham and Veillette 1990; Caron et al. 1992). Thus, in addition to serving as a useful marker for Lck activation, phosphorylation at tyrosine 394 is supposedly instrumental in the up-regulation of the catalytic function of $p56^{lck}$ in vivo.

The tyrosine protein phosphatase(s) dephosphorylating tyrosine 394 in T cells has (have) not been formally identified. Nonetheless, there is experimental evidence indicating that CD45 can catalyze this reaction. First, antibody-mediated co-aggregation of CD4 with CD45 prevents the CD4-induced phosphorylation of $p56^{lck}$ at tyrosine 394 (Ostergaard and Trowbridge 1990). Second, $p56^{lck}$ molecules obtained from CD45-negative T cells exhibit a variable increase in phosphorylation at tyrosine 394, in addition to the above-mentioned increase at tyrosine 505 (Cahir McFarland et al. 1993; Hurley et al. 1993). Considering these data, CD45 may thus have the dual responsibility of dephosphorylating the negative and positive regulatory sites of p56$^{lck}$. Clearly, if both effects are physiologically relevant, they are likely to be differentially regulated at various stages of T cell activation.

## 3.3 Regulation of p56$^{lck}$ by Serine Phosphorylation

Stimulation of T lymphocytes by antibodies against the antigen receptor complex, mitogenic lectins or phorbol esters causes a marked phosphorylation of p56$^{lck}$ at NH2-terminal serine residues (Veillette et al. 1988a, Luo and Sefton 1990b). This phosphorylation is associated with a striking retardation in the electrophoretic mobility of Lck, thereby migrating at 59–60 kDa in sodium dodecyl sulfate-polyacrylamide gels. Through peptide sequencing and site-directed mutagenesis studies, p56$^{lck}$ serine phosphorylation was shown to involve residues 42 and 59 and, to a lesser extent, residues 158 and 194 (Winkler et al. 1993; Watts et al. 1993; Soula et al. 1993). While the kinases effecting these changes in vivo have not been formally identified, in vitro analyses suggest that serine 42 is a target for cyclic AMP-dependent protein kinase and protein kinase C (Winkler et al. 1993). In contrast, phosphorylation at serine 59, which confers the retarded electrophoretic mobility to p56$^{lck}$, can be caused by the mitogen-activated protein (MAP) kinases (Winkler et al. 1993; Watts et al. 1993).

The functional significance of Lck serine phosphorylation is not well understood. Intriguingly, treatment of T cells with phorbol esters or anti-CD3 antibodies also provokes a dissociation of p56$^{lck}$ from CD4, but not from CD8 (Hurley et al. 1989). Hence, serine phosphorylation may change the interactions of Lck with a subset of its regulators. Additionally, since phosphorylation at serine 59 decreases the ability of p56$^{lck}$ to autophosphorylate in vitro (Watts et al. 1993), it is possible that this modification provides an alternative mechanism to repress the catalytic function of p56$^{lck}$ in vivo.

## 3.4 Lymphocyte-Specific Regulation of p56$^{lck}$

In contrast to wild-type Lck molecules isolated from transfected fibroblasts, p56$^{lck}$ polypeptides obtained from T cell lines are poorly phosphorylated at tyrosine 505 (Ostergaard et al. 1989; Cahir McFarland et al. 1993; Hurley et al. 1993; Sieh et al. 1993). As discussed earlier, this hypophosphorylation is seemingly mediated by CD45 and is a requirement for the initiation of TCR-triggered signals (Pingel and Thomas 1989; Koretzky et al. 1990). Intriguingly, these hypophosphorylated or "preactivated" Lck molecules do not cause significant tyrosine protein phosphorylation in resting T cells. This phenomenon is also exemplified by the observation that expression of a constitutively activated Lck mutant (F505 Lck) fails to enhance tyrosine protein phosphorylation in antigen-specific mouse T cell lines prior to TCR stimulation (Abraham et al. 1991; Luo and Sefton 1992). Hence, mechanisms other than tyrosine 505 phosphorylation are clearly involved in the negative regulation of p56$^{lck}$ in resting T lymphocytes.

Recent data indicate that these supplementary inhibitory processes comprise both intramolecular and extramolecular mechanisms (WEIL and VEILLETTE 1994). Indeed, structural-functional studies have demonstrated that the SH3 and SH2 sequences cooperate to repress the catalytic function of $p56^{lck}$ in T cells in the absence of tyrosine 505 phosphorylation. Furthermore, the elevated tyrosine protein phosphatase activity present in T cells is also involved in the negative regulation of $p56^{lck}$, presumably by dephosphorylating the site of Lck autophosphorylation (tyrosine 394). As TCR stimulation permits the biological and biochemical manifestations of activated $p56^{lck}$ molecules in T cells, it is possible that one or more of these inhibitory mechanisms is alleviated by engagement of TCR.

# 4 Functions of $p56^{lck}$ in T Lymphocyte Signal Transduction

## 4.1 Roles in T Cell Activation

### 4.1.1 Involvement of $p56^{lck}$ in CD4- and CD8-Mediated Signal Transduction

An important advance in our understanding of the role of $p56^{lck}$ in normal T cell physiology was provided by the observation that it is physically associated with the CD4 and CD8 T cell surface antigens. Indeed, approximately 50%–90% and 10%–25% of total $p56^{lck}$ molecules are complexed with CD4 and CD8, respectively (VEILLETTE et al. 1988b; RUDD et al. 1988). These interactions have been documented in mature and immature T cells and have been characterized in cells of human, murine and avian origins (VEILLETTE et al. 1988b,1989c; RUDD et al. 1988; VEILLETTE and RATCLIFFE 1991; CHOW et al. 1992). They are dependent on pairs of cysteine residues located in the cytoplasmic domain of CD4 and CD8$\alpha$ and in the unique domain of $p56^{lck}$ (Fig. 4) (SHAW et al. 1989, 1990; TURNER et al. 1990). Whereas the precise biochemical basis for the formation of the CD4-$p56^{lck}$ and CD8-$p56^{lck}$ complexes still needs to be determined, it has been hypothesized to involve a heavy metal ion intermediate.

As discussed by Burakoff and his colleagues in this volume, CD4 and CD8 function as accessory molecules during T cell activation. As a result of their respective ability to bind class II and class I MHC polypeptides on antigen presenting cells, CD4 and CD8 can undergo physical association with the TCR/CD3 complex during antigen stimulation. Hence, both molecules have been termed "coreceptors" of the antigen receptor complex. While early studies provided conflicting data on the roles of CD4 and CD8 in T cell activation, it is now clear that these two glycoproteins enhance TCR-mediated signals. Structure-function analyses have indicated that this property is highly dependent on the ability of CD4 and CD8 to associate with Lck, especially in conditions of sub-optimal antigen stimulation (SLECKMAN et al. 1988; ZAMOYSKA et al. 1989; LETOURNEUR et al. 1990; MICELI et al. 1991; CHALUPNY et al. 1991; GLAICHENHAUS

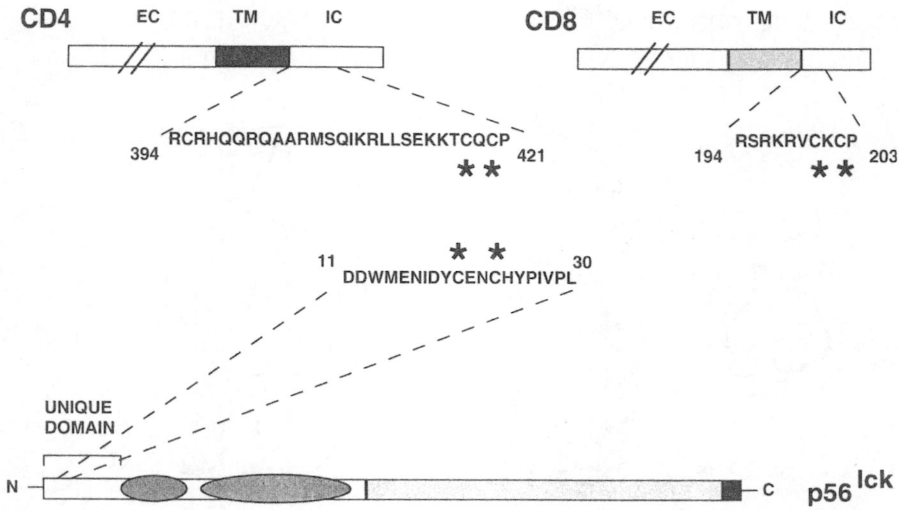


**Fig. 4.** Sequences allowing the interactions between p56$^{lck}$, CD4 and CD8. The sequence motifs allowing p56$^{lck}$ to associate with CD4 and CD8 are shown. The critical cysteine residues are indicated by *asterisks*

et al. 1991). In some experimental systems , however, CD4 and CD8 have been observed to increase T cell responsiveness in the absence of association with p56$^{lck}$ (Miceli et al. 1991; Zerbib et al. 1994). This Lck-independent function is hypothesized to reflect the additional ability of CD4 and CD8 to serve as adhesion molecules, stabilizing physical interactions between the T cell and the antigen presenting cell. Additionally, it may reflect the involvement of these two receptors in alternative pathways of signal transduction (Telfer and Rudd 1991).

A potential mechanism by which CD4-p56$^{lck}$ (and presumably CD8-p56$^{lck}$) potentiates TCR-mediated signal transduction is suggested by the finding that antibody-mediated aggregation of CD4 causes a rapid increase in the catalytic activity of p56$^{lck}$, as measured by immune complex kinase reactions (Veillette et al. 1989b; Luo and Sefton 1990a). Furthermore, engagement of CD4 triggers in vivo phosphorylation of p56$^{lck}$ at tyrosine 394, the site of positive regulation (Veillette et al. 1989a; Luo and Sefton 1990a). Thus, it appears likely that CD4 potentiates TCR signaling both by activating p56$^{lck}$ and by recruiting the activated Lck molecules in the vicinity of the antigen receptor complex (Fig. 5). The recruited Lck molecules would then augment TCR signals by phosphorylating rate-limiting substrates such as CD3, ζ and Zap-70.

An additional means by which CD4-bound Lck can enhance TCR-triggered tyrosine protein phosphorylation was recently suggested by Xu and Littman (1993). These authors observed that the CD4-associated p56$^{lck}$ could improve TCR signaling by a mechanism unrelated to its tyrosine protein kinase activity, but instead relying on its SH2 sequence. Possibly, the SH2 domain of p56$^{lck}$ fulfills this function by recruiting other phosphotyrosine-containing signaling molecules or, alternatively, by protecting tyrosine phosphorylated substrates

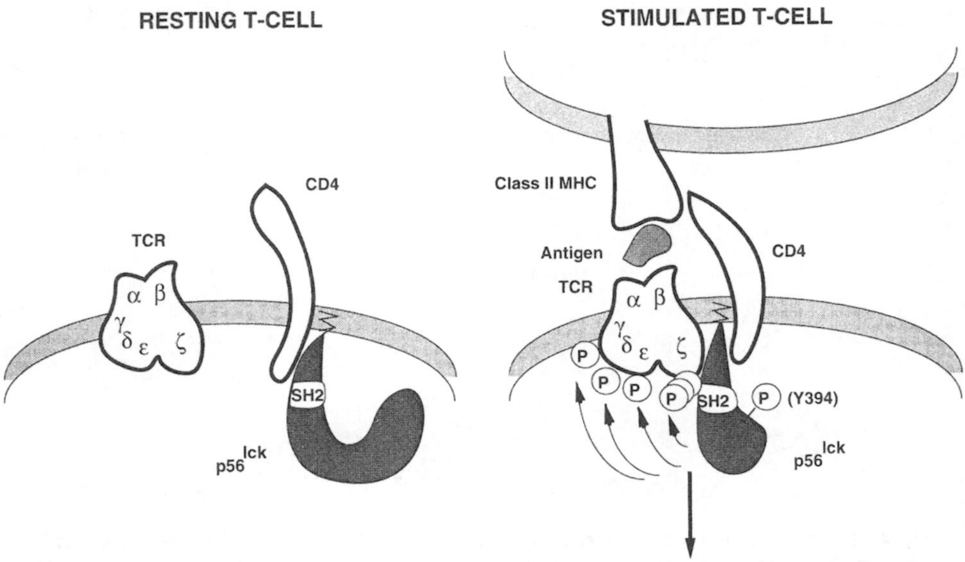


**Fig. 5.** Involvement of p56$^{lck}$ in CD4-mediated signal transduction. Although not shown on this schema, CD4-p56$^{lck}$ is presumed to undergo dimerization or oligomerization upon binding of CD4 to class II MHC molecules. Following dimerization/oligomerization, p56$^{lck}$ becomes activated and shows increased phosphorylation at its positive regulatory site, tyrosine 394. Because class II MHC molecules induce an association of CD4 with TCR, p56$^{lck}$ becomes juxtaposed to the T cell receptor (*TCR*) complex and could potentially phosphorylate rate-limiting targets such as CD3 and ζ. The phosphorylation of these TCR-associated chains allegedly causes a dramatic increase in TCR-induced tyrosine phosphorylation of other cellular proteins

from the action of phosphatases. Furthermore, by binding the tyrosine phosphorylated CD3 and ζ chains, the SH2 sequence may stabilize the association between CD4 and TCR.

### 4.1.2 Involvement of p56$^{lck}$ in T Cell Receptor-Mediated Signal Transduction

There is accumulating evidence that p56$^{lck}$ plays a central role in TCR-mediated signal transduction independently of its association with CD4 and CD8. This view was first suggested by the finding that expression of a constitutively activated version of p56$^{lck}$ (F505 p56$^{lck}$) in a CD4- and CD8-negative mouse T cell hybridoma (BI-141) greatly enhanced antigen receptor-induced tyrosine protein phosphorylation and lymphokine secretion (Abraham et al. 1991). Subsequent characterization of Lck-deficient T cell lines further substantiated this concept. Indeed, Lck-deficient variants of the human T leukemia cell line Jurkat and the mouse cytotoxic T cell line CTLL-2 were shown to have greatly diminished TCR-triggered signals, which could be rescued by gene transfer-mediated expression of wild-type p56$^{lck}$ (Strauss and Weiss 1992; Karnitz et al. 1992). These findings also indicate that other tyrosine protein kinases expressed in T cells (including $p59^{fynT}$) are unable to substitute for Lck in the initiation of TCR signaling.

The exact mechanism by which $p56^{lck}$ mediates TCR signaling in the absence of CD4 and CD8 has not been elucidated. Nevertheless, structure-function analyses demonstrate that this process necessitates the association of $p56^{lck}$ with the plasma membrane, and that it depends on the presence of an intact Lck SH2 domain (CARON et al. 1992). These prerequisites likely imply that the recruitment of Lck in this pathway involves an SH2 domain-mediated association with a tyrosine phosphorylated membrane-bound intermediate (Fig. 6).

Intriguingly, HAUGHN et al. (1992) reported that expression of CD4 in a CD4-negative, class II MHC-restricted antigen-specific T cell clone (5.32.10) markedly diminished the ability of anti-TCR antibodies to stimulate tyrosine protein phosphorylation and cellular proliferation. Notably, however, this inhibitory effect could be overcome by costimulation of TCR and CD4. Contrary to wild-type CD4 molecules, introduction of CD4 polypeptides incapable of associating with $p56^{lck}$ did not down-regulate TCR-mediated signals in this system. These data suggest that the binding of CD4 to $p56^{lck}$ decreases the ability of the tyrosine protein kinase to participate in TCR signaling, unless, of course, TCR is deliberately co-aggregated with CD4. This "sequestration" of $p56^{lck}$ is probably critical for the control of normal immune responsiveness, as it would permit productive T cell activation only when antigens are presented by the appropriate MHC molecules.

### 4.1.3 Potential Involvement of $p56^{lck}$ in Other T Cell Signaling Pathways

In addition to CD4 and CD8, $p56^{lck}$ can form complexes and/or be enzymatically regulated by several other T cell surface molecules. These include the β chain of the IL-2 receptor (HORAK et al. 1991; HATAKEYAMA et al. 1991), CD2 (BELL et al. 1992), CD5 (BURGESS et al. 1992; RAAB et al. 1994), CD45 (SCHRAVEN et al. 1991; KORETZKY et al. 1993) and 4-1BB (KIM et al. 1993). Whereas the pertinence of these findings is unknown, they suggest that Lck may participate in a multitude of other T cell signal transduction responses.

## 4.2 Roles of $p56^{lck}$ in T Cell Development

### 4.2.1 Regulation of T Cell Maturation by $p56^{lck}$

Mature T cells are developed in the thymus from precursors that arise in the bone marrow (reviewed in VON BOEHMER and KISIELOW 1993). During thymic differentiation, T lymphocytes undergo an ordered series of phenotypic changes which are influenced by the specificity of the TCR expressed on individual thymocytes. As discussed elsewhere (VON BOEHMER and KISIELOW 1993), immature thymocytes are subjected to both negative and positive selection. As a consequence, auto-reactive lymphocytes are eliminated and thymocytes bearing "useful" antigen

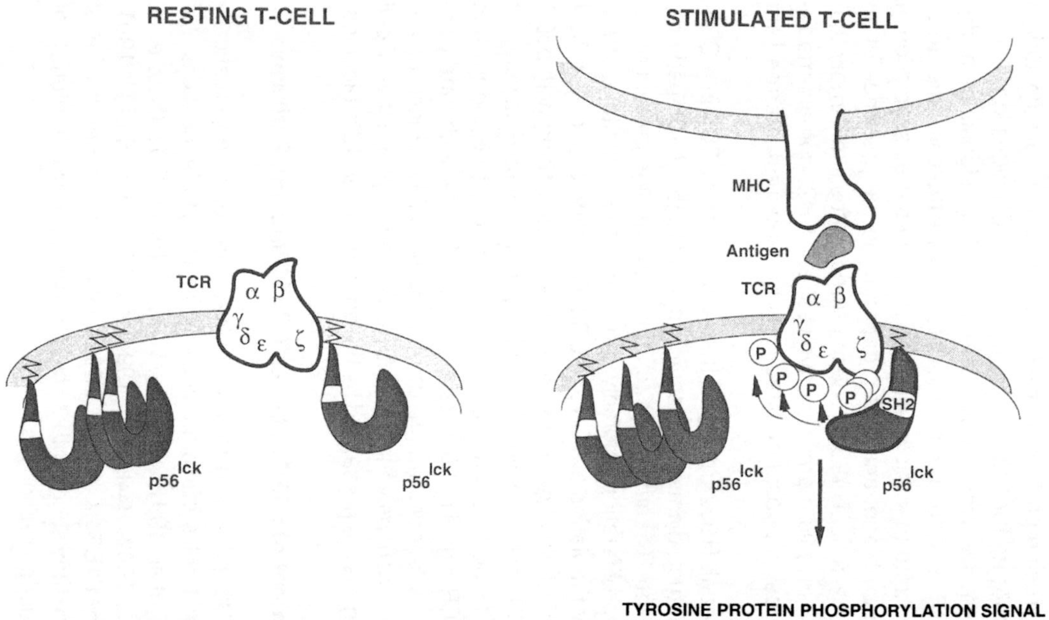


**Fig. 6.** Involvement of $p56^{lck}$ in T cell receptor (*TCR*)-mediated signal transduction. In resting T cells, Lck is maintained in a "preactivated" state, as a result of constitutive dephosphorylation of its negative regulatory site, tyrosine 505, by the tyrosine protein phosphatase CD45. It is postulated that TCR stimulation by MHC-bound antigens provokes a conformational change in the cytoplasmic domains of CD3 and ζ, which become susceptible to phosphorylation by neighbouring $p56^{lck}$ molecules

receptors develop into functional T lymphocytes which emigrate to peripheral lymphoid organs.

Early on, thymocytes do not express CD4, CD8 or TCR. At a subsequent stage, they acquire surface expression of the β chain of TCR, in combination with a 33 kDa glycoprotein (gp33) (GROETTRUP et al. 1993). This complex is hypothesized to transduce intracellular biochemical signals critical for further T cell maturation. Later, thymocytes exhibit expression of both CD4 and CD8 (double-positive thymocyte stage), and plasma membrane-associated TCR αβ heterodimers become detectable. The interaction of TCR αβ with class I or class II MHC-bound antigens leads to differentiation into $CD4^-$ $CD8^+$ or $CD4^+$ $CD8^-$ T cells, respectively (single-positive thymocyte stage). These cells are the immediate precursors of the single-positive T cells found in spleen, lymph nodes and peripheral blood.

The engineering of $p56^{lck}$-deficient mice by homologous recombination in embryonic stem cells clearly demonstrated its importance in T cell maturation (MOLINA et al. 1992). Indeed, animals lacking $p56^{lck}$ exhibit a marked thymic atrophy, which is largely due to a decrease in the number of double-positive and single-positive thymocytes. A related phenotype was observed in transgenic mice expressing a "dominant-negative" variant of Lck (kinase-deficient lysine 273-to-arginine 273 Lck mutant) (LEVIN et al. 1993). Interestingly, these effects of *lck* gene ablation are more dramatic than those caused by combined disruption of the mouse CD4 and CD8 genes, implying that an important component of the thymic function of $p56^{lck}$ is independent of interactions with the two T cell surface antigens. It is tempting to speculate that, in immature thymocytes, $p56^{lck}$ takes part in additional cell surface receptor-triggered signaling pathways (such as that regulated by the TCR β-gp33 complex).

In addition to affecting the development of αβ T cells, the absence of $p56^{lck}$ causes profound perturbations in the development of γδ T cells. Indeed, by breeding $p56^{lck}$-deficient mice with TCR γδ transgenics, PENNINGER et al. (1993) showed that $p56^{lck}$ is also necessary for the development of γδ T lymphocytes in the thymus, although it is dispensable for the phenotypic maturation of both TCR αβ-and TCR γδ-expressing T cells in the intestinal epithelium.

Further to its role in the general process of T cell maturation, $p56^{lck}$ may be specifically involved in the control of allelic exclusion of the TCR β genes. This concept is supported by at least two lines of experimental evidence. First, thymocytes from transgenic mice overexpressing a constitutively activated version of $p56^{lck}$ (F505 Lck) display a specific reduction in Vβ-Dβ, but not Dβ-Jβ, rearrangement (ANDERSON et al. 1992). Second, a "dominant-negative" Lck variant, which presumably interferes with the function of the endogenous wild-type Lck, counteracts the ability of a transgenic TCR β gene to repress Vβ rearrangement (ANDERSON et al. 1993). Even though the precise mechanism by which Lck may regulate allelic exclusion is not defined, it is seemingly unrelated to its association with CD4 or CD8. Once again, this may reflect the participation of $p56^{lck}$ in TCR β-gp33-mediated signals.

### 4.2.2 CD4-p56$^{lck}$ Regulates T Cell Receptor Expression by Controlling Its Sequestration in the Endoplasmic Reticulum

When removed from their thymic microenvironment and incubated at 37°C, double-positive thymocytes show a rapid increase in the cell surface expression of TCR (NAKAYAMA et al. 1990). This induction is due to the release of a pool of TCR complexes previously sequestered and degraded in the endoplasmic reticulum (ER) (BONIFACINO et al. 1990). As anti-CD4 antibody stimulation of ex vivo double-positive thymocytes can reconstitute this ER retention, it is assumed to be the result of CD4-p56$^{lck}$-mediated signals (NAKAYAMA et al. 1990). Consistent with this possibility is the finding that overexpression of CD4 in the thymus diminishes ER retention of TCR and increases TCR levels on double-positive thymocytes (NAKAYAMA et al. 1993). In light of the limiting amounts of p56$^{lck}$, the added CD4 receptors presumably behave as "dominant-negative" molecules, unable to transduce class II-MHC-induced tyrosine protein phosphorylation signals. In contrast to CD4, engagement of CD8 does not cause down-regulation of TCR on CD4$^{+}$ CD8$^{+}$ thymocytes (NAKAYAMA et al. 1990; WIEST et al. 1993). This difference may be a reflection of the significantly lower extent of association of CD8 with p56$^{lck}$ when compared with CD4. While the physiological significance of these various observations is not established, it is conceivable that the relief of TCR sequestration allows CD4-positive cells departing the thymus to acquire a more competent antigen recognition machinery.

## 4.3 Involvement of p56$^{lck}$ in Lymphoid Malignancies

p56$^{lck}$ was first identified as an abundant phosphotyrosine-containing protein in membranes of LSTRA, a Moloney murine leukemia virus (MoMuLV)-induced mouse thymoma cell line (GACON et al. 1982; CASNELLIE et al. 1982; VORONOVA et al. 1985). Further studies showed that the *lck* gene is rearranged in LSTRA, as a consequence of insertion of a MoMuLV long terminal repeat (LTR) in the 5' noncoding portion of the gene (Fig. 2B) (MARTH et al. 1985; VORONOVA and SEFTON 1986; GARVIN et al. 1988; ADLER et al. 1988). This rearrangement causes a seven-fold increase in the rate of *lck* transcription. Furthermore, by disrupting the 5' untranslated region of mature *lck* transcripts, the rearrangement enhances *lck* mRNA translation by seven- to tenfold (MARTH et al. 1988b). Hence, in combination, these modifications allow for a 50-fold increase in the abundance of p56$^{lck}$ in LSTRA cells. A similar *lck* gene rearrangement was detected in Thy-19 (ADLER et al. 1988), another MoMuLV-induced mouse thymoma cell line, and in several MoMuLV-induced rat thymomas (SHIN and STEFFEN 1993).

Direct evidence that augmented expression or activity of p56$^{lck}$ is sufficient to lead to lymphoid transformation was provided by analyses of transgenic mice overexpressing this tyrosine protein kinase (ABRAHAM et al. 1991a,b). Indeed, in addition to showing complex abnormalities of T cell maturation, p56$^{lck}$-overexpressing mice exhibit a high propensity to develop malignant thymomas. The frequency of these tumors is proportional to the extent of Lck overexpression and is greater in mice bearing constitutively activated Lck polypeptides (F505 Lck).

The *lck* gene is also rearranged in at least two human T cell acute lymphoblastic leukemia cell lines, SUP-T12 and HSB-2 (TYCKO et al. 1991; BURNETT et al. 1991). Both cell lines show a translocation between chromosomes 1 and 7, which fuses the 5' elements of the *lck* gene to segments of the TCR β gene. These structural alterations are paralleled by a significant increase in the abundance of $p56^{lck}$ in these leukemic cells. The rearranged *lck* allele in HSB-2 cells also carries several point mutations, which introduce amino acid substitutions in the Lck protein sequence (WRIGHT et al. 1994). Importantly, these mutations cause $p56^{lck}$ to acquire a frank oncogenic potential towards rodent fibroblasts. Combined, these data indicate that changes in the abundance and/or activity of $p56^{lck}$ may play a role in the pathogenesis of human leukemias and lymphomas. Obviously, the identification of structural alterations of the *lck* gene in primary samples of human lymphoid neoplasias will help substantiate this view.

## 5 The Targets of $p56^{lck}$ in T Lymphocytes

Over the past few years, significant progress has been made in identifying the cellular proteins that mediate the effects of Lck in T lymphocytes (Table 1). As one would expect, several of these targets undergo tyrosine phosphorylation upon antibody-mediated aggregation of TCR. However, it should be pointed out that, while TCR-mediated phosphorylation is enhanced by $p56^{lck}$, this functional improvement may not be fully mediated by the direct action of $p56^{lck}$. Possibly, part of these tyrosine protein phosphorylation changes are caused by tyrosine protein kinases regulated by $p56^{lck}$, such as Zap-70.

The CD3 components (γ, δ and ε) and the ζ chains of the TCR complex undergo prompt TCR-triggered tyrosine phosphorylation within the so-called ITAMS (SAMELSON et al. 1986; RETH 1989; QIAN et al. 1993; STRAUSS and WEISS 1993). Phosphorylation at these sites is thought to allow recruitment of SH2 domain-containing proteins, e.g., Zap-70, phospholipase C (PLC)-γ1, Vav and Shc, in TCR-mediated signal transduction.

TCR stimulation also causes tyrosine phosphorylation of the 145 kDa PLC-γ1 (PARK et al. 1991; WEISS et al. 1991). In other systems, tyrosine phosphorylation of PLC-γ1 has been shown to stimulate its catalytic activity and to be responsible for hydrolysis of PI bisphosphate ($PIP_2$) into diacylglycerol (DAG) and inositol trisphosphate ($IP_3$) (NISHIBE et al. 1990). These last two metabolites lead to activation of the serine/threonine-specific protein kinase C and to an increase in intracellular calcium, respectively.

Vav is a 95 kDa polypeptide predicted to possess guanine nucleotide exchange activity towards low molecular weight G-proteins. On the basis of homology with other exchange factors, Vav is expected to act on members of

**Table 1.** Targets of p56[lck] in T lymphocytes

| Substrate | Mass | Functions | Effects of T cell activation |
|---|---|---|---|
| Phospholipase C-γ1 | 145 | Hydrolyses $PIP_2$ in $IP_3$ plus DAG | Tyrosine phosphorylation Activation |
| Vav | 95 | Guanine nucleotide exchange factor (Ras!, Rho?) | Tyrosine Phosphorylation Activation |
| Phosphatidyl inositol 3' kinase | 85+110 | Phosphorylates 3' position inositol ring | Activation |
| Zap-70 | 70 | Tyrosine protein kinase | Tyrosine phosphorylation Association with CD3 and ζ |
| Shc | 46–52 | Adaptor for Grb2-mSOS Activation of Ras | Tyrosine phosphorylation Association with CD3 and ζ (?) |
| CD3, ζ | 16–28 | Orchestrate protein-protein interactions | Tyrosine phosphorylation |

$PIP_2$, phosphatidyl inositol bisphosphate; $IP_3$, inositol trisphosphate, DAG, diacylglycerol

the Rho family. Surprisingly, however, GULBINS et al. (1993) reported that Vav regulates p21[ras]. As the TCR-triggered tyrosine phosphorylation of Vav increases its catalytic activity, it is possible that Vav participates in the activation of small G-proteins during T cell activation (BUSTELO et al. 1992; MARGOLIS et al. 1992; GULBINS et al. 1993). However, direct evidence supporting this view remains to be obtained.

Zap-70 is 70 kDa cytoplasmic tyrosine protein kinase expressed in T cells and NK cells which becomes rapidly tyrosine phosphorylated during T cell activation (CHAN et al. 1991, 1992; WANGE et al. 1992). Interestingly, through its two $NH_2$-terminal SH2 domains, Zap-70 associates with the tyrosine phosphorylated TCR complex (CHAN et al. 1991, 1992; WANGE et al. 1993). Zap-70 is thought to amplify TCR-induced tyrosine protein phosphorylation and to provide signals that are critical for T cell activation. This view is especially supported by the recent determination that a subset of humans with severe T cell immunodeficiency carry inactivating mutations in the *zap-70* gene (ARPAÏA et al. 1994; ELDER et al. 1994; CHAN et al. 1994).

Shc is a 46–52 kDa protein bearing a COOH-terminal SH2 domain. It also undergoes tyrosine phosphorylation during T cell activation and has the ability to associate with the tyrosine phosphorylated ζ chain of TCR (RAVICHANDRAN et al. 1993). Tyrosine phosphorylation of Shc provokes recruitment of the Grb2-mSOS complex to the plasma membrane (EGAN et al. 1993). As mSOS possesses intrinsic guanine nucleotide exchange activity towards Ras, the Shc pathway is probably involved in Ras activation.

TCR stimulation induces tyrosine phosphorylation of other polypeptides of 120, 85, 74 and 36 kDa. While the nature of these products is not known, recent data suggest that the 36 kDa substrate may also be involved in a signal converging on Ras activation (BUDAY et al. 1994).

Finally, during T cell activation, p56[lck] physically associates with the 85 kDa subunit of the PI 3' kinase (VOGEL and FUJITA 1993; PRASAD et al. 1993). This association is mediated by the SH3 region and, to a lesser extent, by the SH2 domain of $p56^{lck}$ and causes an increase in the catalytic activity of PI 3' kinase. While PI 3' kinase phosphorylates the inositol ring at the 3' position, the exact physiological impact of this lipid modification remains to be defined.

*Acknowledgments.* We thank the members of our laboratory, as well as Michael Julius, for critical reading of the manuscript. Work in AV's laboratory is funded by grants from the Medical Research Council of Canada, the National Cancer Institute of Canada and the Cancer Research Society. RW is supported by a post-doctoral fellowship from the Cancer Research Society. AV is a scientist of the Medical Research Council of Canada.

## References

Abraham KM, Levin SD, Marth JD, Forbush KA, Perlmutter RM (1991a) Delayed thymocyte development induced by augmented expression of $p56^{lck}$. J Exp Med 173: 1421–1432

Abraham KM, Levin SD, Marth JD, Forbush KA, Perlmutter RM (1991b) Thymic tumorigenesis induced by overexpression of $p56^{lck}$. Proc Natl Acad Sci USA 88: 3977–3981

Abraham N, Veillette A (1990) Activation of $p56^{lck}$ through mutation of a regulatory carboxy-terminal tyrosine residue requires intact sites of autophosphorylation and myristylation. Mol Cell Biol 10: 5197–5206

Abraham N, Miceli MC, Parnes JR, Veillette A (1991) Enhancement of T-cell responsiveness by the lymphocyte-specific tyrosine protein kinase $p56^{lck}$. Nature 350: 62–66

Adler HT, Reynolds PJ, Kelley CM, Sefton BM (1988) Transcriptional activation of *lck* by retrovirus promoter insertion between two lymphoid-specific promoters. J Virol 62: 4113–4122

Amrein KE, Sefton BM (1988) Mutation of a site of tyrosine phosphorylation in the lymphocyte-specific tyrosine protein kinase, $p56^{lck}$, reveals its oncogenic potential in fibroblasts. Proc Natl Acad Sci USA 85: 4247–4251

Amrein KE, Panholzer B, Flint NA, Bannwarth W, Burn P (1993) The Src homology 2 domain of the protein-tyrosine kinase $p56^{lck}$ mediates both intermolecular and intramolecular interactions. Proc Natl Acad Sci USA 90: 10285–10289

Anderson SJ, Abraham KM, Nakayama T, Singer A, Perlmutter RM (1992) Inhibition of T-cell receptor β chain gene rearrangement by overexpression of non-receptor protein tyrosine kinase $p56^{lck}$ EMBO J 13: 4877–4886

Anderson SJ, Levin SD, Perlmutter RM (1993) Protein tyrosine kinase $p56^{lck}$ controls allelic exclusion of T-cell receptor β-chain genes. Nature 365: 552–554

Arpaïa E, Shahar M, Dadi H, Cohen A, Roifman CM (1994) Defective T cel receptor signaling and $CD8^+$ thymic selection in humans lacking Zap-70 kinase. Cell 76: 947–958

Bell GM, Bolen JB, Imboden JB (1992) Association of Src-like protein tyrosine kinases with the CD2 cell surface molecule in rat T lymphocytes and natural killer cells. Mol Cell Biol 12: 5548–5554

Bennett BD, Cowley S, Jiang S, London R, Deng B, Grabarek J, Groopman JE, Goeddel DV, Avraham H (1994) Identification and characterization of novel tyrosine kinase from megakaryocytes. J Biol Chem 369: 1068–1074

Bergman M, Mustelin T, Oetken C, Partanen J, Flint NA, Amrein KE, Autero M, Burn P, Alitalo K (1992) The human $p50^{csk}$ tyrosine kinase phosphorylates $p56^{lck}$ at tyr-505 and down regulates its catalytic activity. EMBO J 11: 2919–2924

Bonifacino JS, McCarthy SA, Maguire JE, Nakayama T, Singer DS, Klausner RD, Singer A (1990) Novel post-translational regulation of TCR expression in $CD4^+CD8^+$ thymocytes influenced by CD4. Nature 344: 247–251

Bräuninger A, Karn T, Strebhardt K, Rübsamen-Waigmann (1993) Characterization of the human CSK locus. Oncogene 8: 1365–1369

Buday L, Egan SE, Viciana PR, Cantrell DA, Downward J (1994) A complex of Grb2 adaptor protein, Sos exchange factor, and a 36-kDa membrane-bound tyrosine phosphoprotein is implicated in Ras activation in T cells. J Biol Chem 269: 9091–9023

Burgess KE, Yamamoto M, Prasad KV, Rudd CE (1992) CD5 acts as a tyrosine kinase substrate within a receptor complex comprising T-cell receptor ζ chain/CD3 and protein-tyrosine kinases p56^lck and p59^fyn. Proc Natl Acad Sci USA 8: 9311–9315

Burnett RC, David JC, Hardan AM, Le Beau MM, Rowley JD, Diaz MO (1991) The *lck* gene is involved in the t(1; 7)(p34; q34) in the T-cell acute lymphoblastic leukemia derived cell line HSB-2. Genes Chromosomes Cancer 3: 461–467

Bustelo XR, Ledbetter JA, Barbacid M (1992) Product of the vav proto-oncogene defines a new class of tyrosine protein kinase substrates. Nature 356: 68–71

Cahir McFarland ED, Hurley TR, Pingel JT, Sefton BM, Shaw A, Thomas ML (1993) Correlation between Src family member regulation by the protein-tyrosine-phosphatase CD45 and transmembrane signaling through the T-cell receptor. Proc Natl Acad Sci USA 90: 1402–1406

Cantley LC, Auger KR, Carpenter C, Duckworth A (1991) Oncogenes and signal transduction. Cell 64: 281–302

Caron L, Abraham N, Pawson T, Veillette A (1992) Structural requirements for enhancement of T-cell responsiveness by the lymphocyte-specific tyrosine protein kinase p56^lck. Mol Cell Biol 6: 2720–2729

Casnellie JE, Harrison ML, Hellstrom KE, Krebs EG (1982) A lymphoma cell line expressing elevated levels of tyrosine protein kinase activity. J Biol Chem 258: 10738–10742

Chalupny NJ, Ledbetter JA, Kavathas P (1991) Association of CD8 with p56^lck is required for early T cell signaling events. EMBO J 5: 1201–1207

Chan AC, Irving BA, Fraser JD, Weiss A (1991) The ζ chain is associated with a tyrosine kinase and upon T-cell antigen receptor stimulation associates with ZAP-70, a 70-kDa tyrosine phosphoprotein. Proc Natl Acad Sci USA 88: 9166–9170

Chan AC, Iwashima M, Turck CW, Weiss A (1992) ZAP-70: a 70 kd protein-tyrosine kinase that associates with the TCR ζ chain. Cell 71: 649–662

Chan AC, Kadlecek TA, Elder ME, Filipovich AH, Kuo WL, Iwashima M, Parslow TG, Weiss A (1994) ZAP-70 deficiency in an autosomal recessive form of severe combined immunodeficiency. Science 264: 1599–1601

Chow LML, Ratcliffe MJH, Veillette A (1992) tkl is the avian homolog of the mammalian lck tyrosine protein kinase gene. Mol Cell Biol 12: 1226–1233

Chow LML, Fournel M, Davidson D, Veillette A (1993) Negative regulation of T-cell receptor signaling by tyrosine protein kinase p50^csk. Nature 365: 156–160

Chow LML, Jarvis C, Hu Q, Nye SH, Gervais FG, Veillette A, Matis LA (1994) Ntk: a Csk-related protein-tyrosine kinase expressed in brain and T-lymphocytes. Proc Natl Acad Sci USA 91: 4975–4979

Cooper JA (1990) The src family of protein-tyrosine kinases, p.85–113. In B. E. Kemp (ed), Peptides and protein phosphorylation. CRC Press, Boca Raton

Deans JP, Kanner SB, Torres RM, Ledbetter JA (1992) Interaction of CD4: lck with the T cell receptor/CD3 complex induces early signaling events in the absence of CD45 tyrosine phosphatase. Eur J Immunol 22: 661–668

Egan SE, Giddings BW, Brooks MW, Buday L, Sizeland AM, Weinberg RA (1993) Association of Sos Ras exchange protein with Grb2 is implicated in tyrosine kinase signal transduction and transformation. Nature 363: 45–51

Einspahr KJ, Abraham RT, Dick CJ, Leibson PJ (1990) Protein tyrosine phosphorylation and p56^lck modification in IL-2 or phorbol ester-activated human natural killer cells. J Immunol 145: 1490–1497

Elder ME, Lin D, Clever J, Chan AC, Hope TJ, Weiss A, Parslow TG (1994) Human severe combined immunodeficiency due to a defect in ZAP-70, a T cell tyrosine kinase. Science 264: 1596–1599

Gacon G, Gisselbrecht S, Piau JP, Boissel JP, Tolle J, Fischer S (1982) High levels of tyrosine protein kinase in a murine lymphoma cell line induced by the Moloney leukemia virus. EMBO J 1: 1579–1582

Garvin AM, Pawar S, Marth JD, Perlmutter RM (1988) Structure of the murine *lck* gene and its rearrangement in a murine lymphoma cell line. Mol Cell Biol 8: 3058–3064

Gervais FG, Chow LML, Lee JM, Branton PE, Veillette A (1993) The SH2 domain is required for stable phosphorylation of p56^lck at tyrosine 505, the negative regulatory site. Mol Cell Biol 13: 7112–7121

Glaichenhaus N, Shastri N, Littman DR, Turner JM (1991) Requirement for association of p56^lck with CD4 in antigen-specific signal transduction in T cells. Cell 64: 511–520

Groettrup M, Ungewiss K, Azogui O, Palacios R, Owen MJ, Hayday AC, von Boehmer H (1993) A novel disulfide-linked heterodimer on pre-T cells consists of the T cell receptor β chain and a 33 kd glycoprotein. Cell 75: 283–294

Gulbins E, Coggeshall KM, Baier G, Katzav S, Burn P, Altman A (1993) Tyrosine kinase-stimulated guanine nucleotide exchange activity of Vav in T cell activation. Science 260: 822–825
Hatakeyama M, Kono T, Kobayashi N, Kawahara A, Levin SD, Perlmutter RM, Taniguchi T (1991) Interaction of the IL-2 receptor with the src-family kinase p56lck: identification of novel intermolecular association. Science 252: 1523–1528
Haughn L, Gratton S, Caron L, Sekaly RP, Veillette A, Julius M (1992) Association of tyrosine kinase p56lck with CD4 inhibits the induction of growth through the $\alpha\beta$ T-cell receptor. Nature 358: 328–331
Horak ID, Gress RE, Lucas PJ, Horak EM, Waldmann TA, Bolen JB (1991) T-lymphocyte interleukin 2-dependent tyrosine protein kinase signal transduction involves the activation of p56lck. Proc Natl Acad Sci USA 88: 1996–2000
Hurley TR, Luo K, Sefton BM (1989) Activators of protein Kinase C induce dissociation of CD4, but not CD8, from p56lck. Science 245: 407–409
Hurley TR, Hyman R, Sefton BM (1993) Differential effects of expression of the CD45 tyrosine protein phosphatase on the tyrosine phosphorylation of the Lck, Fyn, and c-Src tyrosine protein kinases. Mol Cell Biol 13: 1651–1656
June CH, Fletcher MC, Ledbetter JA, Schieven GL, Siegel JN, Philipps AF, Samelson LE (1990) Inhibition of tyrosine phosphorylation prevents T-cell receptor-mediated signal transduction. Proc Natl Acad Sci USA 87: 7722–7726
Karnitz L, Sutor SL, Torigoe T, Reed JC, Bell MP, McKean DJ, Leibson PJ, Abraham RT (1992) Effects of p56lck deficiency on the growth and cytolytic effector function of an interleukin-2-dependent cytotoxic T-cell line. Mol Cell Biol 10: 4521–4530
Kim YJ, Pollok KE, Zhou Z, Shaw A, Bolen JB, Fraser M, Kwon BS (1993) Novel T cell antigen 4-1BB associates with the protein kinase p56lck. J Immunol 151: 1255–1262
Klages S, Adam D, Class K, Fargnoli J, Bolen JB, Penhallow RC (1994) Ctk: a protein-tyrosine kinase related to Csk that defines an enzyme family. Proc Natl Acad Sci USA 91: 2597–2601
Koch CA, Anderson D, Moran MF, Ellis C, Pawson T (1991) SH2 and SH3 domains: elements that control interactions of cytoplasmic signaling proteins. Science 25: 668–674
Koretzky GA, Picus J, Thomas ML, Weiss A (1990) A tyrosine phosphatase CD45 is essential for coupling T-cell antigen receptor to the phosphatidyl inositol pathway. Nature 346: 66–68
Koretzky GA, Kohmetscher M, Ross S (1993) CD45-associated kinase activity requires Lck but not T cell receptor expression in Jurkat T cell line. J Biol Chem 268: 8958–8964
Letourneur F, Gabert J, Cosson P, Blanc D, Davoust J, Malissen B (1990) A signaling role for the cytoplasmic segment of the CD8 $\alpha$ chain detected under limiting stimulatory conditions. Proc Natl Acad Sci USA 87: 2339–2343
Levin SD, Anderson SJ, Forbush KA, Perlmutter RM (1993) Distinct intracellular localisation of Lck and Fyn protein tyrosine kinases in human T lymphocytes. J Cell Biol 125: 639–649
Ley SC, Marsh M, Bebbington CR, Proudfoot K, Jordan P (1994) Distinct intracellular localisation of Lck and Fyn protein tyrosine kinases in human T lymphocytes. J Cell Biol 125: 639–649
Luo K, Sefton BM (1990a) Cross-linking of T cell surface molecules CD4 and CD8 stimulates phosphorylation of the Lck tyrosine protein kinase at the autophosphorylation site. Mol Cell Biol 10: 5305–5313
Luo K, Sefton BM (1990b) Analysis of the sites in p56lck whose phosphorylation is induced by tetradecanoyl phorbol acetate. Oncogene 5: 803–808
Luo K, Sefton BM (1992) Activated Lck tyrosine protein kinase stimulates antigen-independent interleukin-2 production in T cells. Mol Cell Biol 12: 4724–4732
Margolis B, Hu P, Katzav S, Li W, Oliver JM, Ullrich A, Weiss A, Schlessinger J (1992) Tyrosine phosphorylation of vav proto-oncogene product containing SH2 domain and transcription factor motifs. Nature 356: 71–74
Marth JD, Peet R, Krebs EG, Perlmutter RM (1985) A lymphocyte-specific protein tyrosine kinase gene is rearranged and overexpressed in the murine T cell lymphoma LSTRA. Cell 43: 393–404
Marth JD, Cooper JA, King CS, Ziegler SF, Tinker DA, Overell RW, Krebs EG, Perlmutter RM (1988a) Neoplastic transformation induced by an activated lymphocyte-specific protein tyrosine kinase (pp56lck). Mol Cell Biol 8: 540-550
Marth JD, Overell RW, Meier KE, Krebs EG, Perlmutter RM (1988b) Translational activation of the *lck* proto-oncogene. Nature 332: 171–173
McVicar DW, Lal BK, Lloyd A, Kawamura M, Chen YQ, Zhang X, Staples JE, Ortaldo JR, O'Shea JJ (1994) Molecular cloning of lsk, a carboxyl-terminal src kinase (csk) related gene, expressed in leukocytes. Oncogene 9: 2037–2044
Miceli MC, von Hoegen P, Parnes JR (1991) Adhesion versus coreceptor function of CD4 and CD8: role of the cytoplasmic tail in coreceptor activity. Proc Natl Acad Sci USA 88: 2623–2627

Molina TJ, Kishihara K, Siderovski DP, van Ewijk W, Narendran A, Timms E, Wakeham A, Paige C, Hartmann KU, Veillette A, Davidson D, Mak TW (1992) Profound block in thymocyte development in mice lacking p56$^{lck}$. Nature 357: 161–164

Mustelin T, Coggeshall KM, Isakov N, Altman A (1990) T cell antigen receptor-mediated activation of phospholipase C requires tyrosine phosphorylation. Science 247: 1584–1587

Nada S, Okada M, MacAuley A, Cooper JA, Nakagawa H (1991) Cloning of a complementary DNA for a protein-tyrosine kinase that specifically phosphorylates a negative regulatory site of p60$^{c\text{-}src}$. Nature 351: 69–72

Nakayama T, June CH, Munitz TI, Sheard M, McCarthy SA, Sharrow SO, Samelson LE, Singer A (1990) Inhibition of T cell receptor expression and function in immature CD4$^+$ CD8$^+$ cells by CD4. Science 249: 1558–1561

Nakayama T, Wiest DL, Abraham KM, Munitz TI, Perlmutter RM, Singer A (1993) Decreased signaling competence as a result of receptor overexpression of CD4 reduces its ability to activate p56$^{lck}$ tyrosine kinase and to regulate T-cell antigen receptor expression in immature CD4$^+$ CD8$^+$ thymocytes. Proc Natl Acad Sci USA 90: 10534–10538

Nishibe S, Wahl MI, Hernández-Sotomayor SMT, Tonks NK, Rhee SG, Carpenter G (1990) Increase of the catalytic activity of phospholipase C-$\gamma$1 by tyrosine phosphorylation. Science 250: 1253–1256

Oetken C, Couture C, Bergman M, Bonnefoy-Berard N, Williams S, Alitalo K, Burn P, Mustelin T (1994) TCR/CD3-triggering causes increased activity of the p50$^{csk}$ tyrosine kinase and engagement of its SH2 domain. Oncogene 9: 1625–1631

Okada M, Nada S, Yamanashi Y, Yamamoto T, Nakagawa H (1991) CSK: a protein-tyrosine kinase involved in regulation of src family kinases. J Biol Chem 266: 24249–24252

Ostergaard HL, Trowbridge IS (1990) Coclustering CD45 with CD4 or CD8 alters the phosphorylation and kinase activity of p56$^{lck}$. J Exp Med 172: 347–350

Ostergaard HL, Shackelford DA, Hurley TR, Johnson P, Hyman R, Sefton BM, Trowbridge IS (1989) Expression of CD45 alters phosphorylation of the Lck-encoded tyrosine protein kinase in murine lymphoma T-cell lines. Proc Natl Acad Sci USA 86: 8959–8963

Park DJ, Rho YW, Rhee SG (1991) CD3 stimulation causes phosphorylation of phospholipase C on serine and tyrosine residues in a human T-cell line. Proc Natl Acad Sci USA 88: 5453–5456

Partanen J, Armstrong E, Bergman M, Mäkelä TP, Hirvonen H, Huebner K, Alitalo K (1991) cyl encodes a putative cytoplasmic tyrosine kinase lacking the conserved autophosphorylation site (Y416$^{src}$). Oncogene 6: 2013–2018

Pawson T, Gish GD (1992) SH2 and SH3 domains: from structure to function. Cell 71: 359–362

Payne G, Shoelson SE, Gish GD, Pawson T, Walsh CT (1993) Kinetics of p56$^{lck}$ and p60$^{src}$ Src homology 2 domain binding to tyrosine-phosphorylated peptides determined by a competition assay or surface plasmon resonance. Proc Natl Acad Sci USA 90: 4902–4906

Penninger J, Kishihara K, Molina T, Wallace VA, Timms E, Hedrick SM, Mak TW (1993) Requirement for tyrosine kinase p56$^{lck}$ for thymic development of transgenic T cells. Science 260: 358–361

Peri KG, Gervais FG, Weil R, Davidson D, Gish GD, Veillette A (1993) Interactions of the SH2 domain of lymphocyte-specific tyrosine protein kinase p56$^{lck}$ with phosphotyrosine-containing proteins. Oncogene 8: 2765–2772

Perlmutter RM, Levin SD, Appleby MW, Anderson SJ, Alberola-Ila J (1993) Regulation of lymphocytes function by protein phosphorylation. Annu Rev Biochem 11: 451–499

Pingel JT, Thomas ML (1989) Evidence that the leukocyte-common antigen is required for antigen-induced T lymphocyte proliferation. Cell 58: 1055–1065

Prasad KVS, Kapeller R, Janssen O, Repke H, Duke-Cohan JS, Cantley LC, Rudd CE (1993) Phosphatidylinositol (PI) 3-kinase and PI 4-kinase binding to the CD4-p56$^{lck}$ complex: the p56$^{lck}$ SH3 domain binds to PI 3-kinase but not PI 4-kinase. Mol Cell Biol 13: 7708–7717

Qian D, Griswold-Prenner I, Rosner MR, Fitch FW (1993) Multiple components of the T cell antigen receptor complex become tyrosine-phosphorylated upon activation. J Biol Chem 268:4488–4493

Raab M, Yamamoto M, Rudd CE (1994) The T-cell antigen CD5 acts as a receptor and substrate for the protein-tyrosine kinase p56$^{lck}$. Mol Cell Biol 14: 2862–2870

Ravichandran KS, Lee KK, Songyang Z, Cantley LC, Burn P, Burakoff SJ (1993) Interaction of Shc with the $\zeta$ chain of the T cell receptor upon T cell activation. Science 262: 902–904

Reth M (1989) Antigen receptor tail clue. Nature 338: 383

Reynolds PJ, Lesley J, Trotter J, Schulte R, Hyman R, Sefton BM (1990) Changes in the relative abundance of type I and type II lck mRNA transcripts suggest differential promoter usage during T-cell development. Mol Cell Biol 10: 4266–4270

Reynolds PJ, Hurley TR, Sefton BM (1992) Functional analysis of the SH2 and SH3 domains of the Lck tyrosine protein kinase. Oncogene 7: 1949–1956

Rudd CE, Trevillyan JM, Dasgupta JD, Wong LL, Schlossman SF (1988) The CD4 receptor is complexed in detergent lysates to a protein-tyrosine kinase (pp58) from human T lymphocytes. Proc Natl Acad Sci USA 85: 5190-5194

Ruzzene M, James P, Brunati AM, Donella-Deana A, Pinna LA (1994) Regulation of c-Fgr protein kinase by c-Src kinase (CSK) and by polycationic effectors. J Biol Chem 269: 15885–15891

Sabe H, Okada M, Nakagawa H, Hanafusa H (1992) Activation of c-Src in cells bearing v-Crk and its suppression by Csk. Mol Cell Biol 12: 4706–4713

Sakano S, Iwama A, Inazawa J, Ariyama T, Ohno M, Suda T (1994) Molecular cloning of a novel non-receptor tyrosine kinase, HYL (hematopoietic consensus tyrosine-lacking kinase). Oncogene 9: 1155–1161

Samelson LE, Klausner RD (1992) Tyrosine kinases and tyrosine-based activation motifs. Current research on activation via the T cell antigen receptor. J Biol Chem 267: 24913–24916

Samelson LE, Patel MD, Weissman AM, Harford JB, Klausner RD (1986) Antigen activation of murine T cells induces tyrosine phosphorylation of a polypeptide associated with the antigen receptor. Cell 46: 1083–1090

Schraven B, Kirchgessner H, Gaber B, Samstag Y, Meuer S (1991) A functional complex is formed in human T lymphocytes between the protein tyrosine phosphatase CD45, the protein tyrosine kinase $p56^{lck}$ and pp32, a possible common substrate. Eur J Immunol 10: 2469–2477

Shaw AS, Amrein KE, Hammond C, Stern DF, Sefton BM, Rose JK (1989) The Lck tyrosine protein kinase interacts with the cytoplasmic tail of the CD4 glycoprotein through its unique amino-terminal domain. Cell 59: 627–636

Shaw AS, Chalupny J, Whitney JA, Hammond C, Amrein KE, Kavathas P, Sefton BM, Rose JK (1990) Short related sequences in the cytoplasmic domains of CD4 and CD8 mediate binding to the amino-terminal domain of the $p56^{lck}$ tyrosine protein kinase. Mol Cell Biol 10: 1853–1862

Shenoy-Scaria A, Timson Gauen LK, Kwong J, Shaw A, Lublin DM (1993) Palmitylation of an amino-terminal cysteine motif of protein tyrosine kinase $p56^{lck}$ and $p59^{fyn}$ mediates interaction with glycosyl-phosphatidylinositol-anchored proteins. Mol Cell Biol 13: 6385–6392

Shin S, Steffen DL (1993) Frequent activation of the *lck* gene by promoter insertion and aberrant splicing in murine leukemia virus-induced rat lymphomas. Oncogene 8: 141–149

Shiroo M, Goff L, Biffen M, Shivnan E, Alexander D (1992) CD45 tyrosine phosphatase-activated $p59^{fyn}$ couples T cell antigen receptor to pathways of diacylglycerol production, protein kinase C activation and calcium influx. EMBO J 11: 4887–4897

Sieh M, Bolen JB, Weiss A (1993) CD45 specifically modulates binding of Lck to a phosphopeptide encompassing the negative regulatory tyrosine of Lck. EMBO J 12: 315–321

Sleckman BP, Peterson A, Foran JA, Gorga JC, Kara CJ, Strominger JL, Burakoff SJ, Greenstein JL (1988) Functional analysis of cytoplasmic domain-deleted mutant of the CD4 molecule. J Immunol 141: 49–54

Soula M, Rothhut B, Camoin L, Guillaume JL, Strosberg D, Vorherr T, Burn P, Meggio F, Fischer S, Fagard R (1993) Anti-CD3 and phorbol ester induce distinct phosphorylated sites in the SH2 domain of $p56^{lck}$. J Biol Chem 268: 27420–27427

Strauss DB, Weiss A (1992) Genetic evidence for the involvement of the Lck tyrosine kinase in signal transduction through the T cell antigen receptor. Cell 70: 585–593

Strauss DB, Weiss A (1993) The CD3 chains of the T cell antigen receptor associate with the ZAP-70 tyrosine kinase and are tyrosine phosphorylated after receptor stimulation. J Exp Med 178: 1523–1530

Taieb J, Vitte-Mony I, Auffredou MT, Dorseuil O, Gacon G, Bertoglio J, Vazquez A (1993) Regulation of $p56^{lck}$ kinase expression and control of DNA synthesis in activated human B lymphocytes. J Biol Chem 268: 9169–9171

Telfer JC, Rudd CE (1991) A 32-kD GTP-binding protein associated with the CD4-$p56^{lck}$ and CD8-$p56^{lck}$ T cell receptor complexes. Science 254: 439–441

Trowbridge IS, Thomas ML (1994) CD45: an emerging role as a protein tyrosine phosphatase required for lymphocyte activation and development. Annu Rev Immunol 12: 85–116

Turner JM, Brodsky MH, Irving BA, Levin SD, Perlmutter RM, Littman DR (1990) Interaction of the unique N-terminal region of tyrosine kinase $p56^{lck}$ with cytoplasmic domains of CD4 and CD8 is mediated by cysteine motifs. Cell 60: 755–765

Tycko B, Smith SD, Sklar J (1991) Chromosomal translocation joining *lck* and TCR β loci in human T cell leukemia. J Exp Med 174: 867–873

Veillette A, Davidson D (1992) Src-related protein tyrosine kinases and T-cell receptor signaling. Trends Genet 8: 61–66

Veillette A, Ratcliffe MJH (1991) Avian CD4 and CD8 interact with a cellular tyrosine protein kinase homologous to mammalian p56$^{lck}$. Eur J Immunol 21: 397–401

Veillette A, Foss FM, Sausville EA, Bolen JB, Rosen N (1987) Expression of the lck tyrosine kinase gene in human colon carcinoma and other non-lymphoid human tumor cell lines. Oncogene Res 1: 357–374

Veillette A, Horak ID, Bolen JB (1988a) Post-translational alterations of the tyrosine kinase p56$^{lck}$ in response to activators of protein kinase C. Oncogene Res 2: 385–401

Veillette A, Bookman MA, Horak EM, Bolen JB (1988b) The CD4 and CD8 T cell surface antigens are associated with the internal membrane tyrosine-protein kinase p56$^{lck}$. Cell 55: 301–308

Veillette A, Horak ID, Horak EM, Bookman MA, Bolen JB (1988c) Alterations of the lymphocyte-specific protein tyrosine kinase (p56$^{lck}$) during T-cell activation. Mol Cell Biol 8: 4353–4361

Veillette A, Bolen JB, Bookman MA (1989a) Alterations in tyrosine protein phosphorylation induced by antibody-mediated cross-linking of the CD4 receptor on T-lymphocytes. Mol Cell Biol 9: 4441–4446

Veillette A, Bookman MA, Horak EM, Samelson LE, Bolen JB (1989b) Signal transduction through the CD4 receptor involves the activation of the internal membrane tyrosine-protein kinase p56$^{lck}$. Nature 338: 257–259

Veillette A, Zuniga-Pflucker JC, Bolen JB, Kruisbeek AM (1989c) Engagement of CD4 and CD8 expressed on immature thymocytes induces activation of intracellular tyrosine phosphorylation pathways. J Exp Med 170: 1671–1680

Veillette A, Caron L, Fournel M, Pawson T (1992) Regulation of the enzymatic function of the lymphocyte-specific tyrosine protein kinase p56$^{lck}$ by the non-catalytic SH2 and SH3 domains. Oncogene 7: 971–980

Volarevic S, Niklinska BB, Burns CM, Yamada H, June CH, Dumont FJ, Ashwell JD (1992) The CD45 tyrosine phosphatase regulates phosphotyrosine homeostasis and its loss reveals a novel pattern of late T cell receptor-induced $Ca^{2+}$ oscillations. J Exp Med 176: 835–844

Vogel LB, Fujita DJ (1993) The SH3 domain of p56$^{lck}$ is involved in binding to phosphatidylinositol 3'-kinase from T lymphocytes. Mol Cell Biol 13: 7408–7417

von Boehmer H, Kisielow P (1993) Lymphocyte lineage commitment: instruction versus selection. Cell 73: 207–208

Voronova AF, Sefton BM (1986) Expression of a new tyrosine protein kinase is stimulated by retrovirus promoter insertion. Nature 319: 682–685

Voronova AF, Buss JE, Patchinsky T, Hunter T, Sefton BM (1985) Characterization of the protein apparently responsible for the elevated tyrosine protein kinase activity in LSTRA cells. Mol Cell Biol 4: 2705–2713

Wange RL, Tony Kong AN, Samelson LE (1992) A tyrosine-phosphorylated 70-kDa protein binds a photoaffinity analogue of ATP and associates with both the zeta chain and CD3 components of the activated T cell antigen receptor. J Biol Chem 267: 11685–11688

Wange RL, Malek SN, Desiderio S, Samelson LE (1993) Tandem SH2 domains of ZAP-70 bind to T cell antigen receptor ζ and CD3ε from activated Jurkat T cells. J Biol Chem 268: 19797–19801

Watts JDM, Sanghera JS, Pelech SL, Aebersold R (1993) Phosphorylation of serine 59 of p56$^{lck}$ in activated T cells. J Biol Chem 258: 23275–23282

Weaver CT, Pingel JT, Nelson JO, Thomas ML (1991) CD8$^{+}$ T-cell clones deficient in the expression of the CD45 protein tyrosine phosphatase have impaired responses to T-cell receptor stimuli. Mol Cell Biol 11: 4415–4422

Weil R, Veillette A (1994) Intramolecular and extramolecular mechanisms repress the catalytic function of p56$^{lck}$ in resting T-lymphocytes. J Biol Chem 269: 22830–22838

Weiss A, Littman DR (1994) Signal transduction by lymphocyte antigen receptors. Cell 76: 263–294

Weiss A, Koretzky G, Schatzman RC, Kadlecek T (1991) Functional activation of the T-cell antigen receptor induces tyrosine phosphorylation of phospholipase C-γ 1. Proc Natl Acad Sci USA 88: 5484–5488

Wiest DL, Yuan L, Jefferson J, Benveniste P, Tskos M, Klausner RD, Glimcher LH, Samelson LE, Singer A (1993) Regulation of T cell receptor expression in immature CD4$^{+}$ CD8$^{+}$ thymocytes by p56$^{lck}$ tyrosine kinase: basis for differential signaling by CD4 and CD8 in immature thymocytes expressing both coreceptor molecules. J Exp Med 178: 1701–1712

Wildin RS, Garvin AM, Pawar S, Lewis DB, Abraham KM, Forbush KA, Ziegler SF, Allen JM, Perlmutter RM (1991) Developmental regulation of *lck* gene expression in T lymphocytes. J Exp Med 173: 383–393

Winkler DG, Park I, Kim TU, Payne NS, Walsh CT, Strominger JL, Shin J (1993) Phosphorylation of Ser-42 and Ser-59 in the N-terminal region of the tyrosine kinase $p56^{lck}$. Proc Natl Acad Sci USA 90: 5176–5180

Wright DD, Sefton BM, Kamps MP (1994) Oncogenic activation of the Lck protein accompanies translocation of the *lck* gene in the human HSB-2 T-cell leukemia. Mol Cell Biol 14: 2429–2437

Xu H, Littman DR (1993) A kinase-independent function of Lck in potentiating antigen-specific T-cell activation. Cell 74: 633–643

Zamoyska R, Derham P, Gorman SD, von Hoegen P, Bolen JB, Veillette A, Parnes JR (1989) Inability of CD8 $\alpha'$ polypeptides to associate with $p56^{lck}$ correlates with impaired function in vitro and lack of expression in vivo. Nature 342: 278–281

Zerbib AC, Reske-Kunz AB, Lock P, Sekaly RP (1994) CD4-mediated enhancement or inhibition of T cell activation does not require the CD4: $p56^{lck}$ association. J Exp Med 179: 1973–1983

# The Regulation and Function of the CD4 Coreceptor During T Lymphocyte Development

N. KILLEEN[1] and D.R. LITTMAN[2]



## 1 Introduction

The progression of thymocytes through sequential developmental stages is marked by an ordered pattern of expression of many different cell surface molecules (KRUISBEEK 1993). Of these, the CD4 and CD8 glycoproteins are noteworthy because their differential expression defines the two major lineages of mature peripheral T cells (SWAIN 1983). Although mature cells express one molecule or the other, the large majority of immature thymocytes expresses both CD4 and CD8. These double positive (DP) cells are subjected to a stringent selection process which favors survival of the few cells that can interact productively with host MHC molecules (HOGQUIST et al 1994; VON BOEHMER 1994). Down-regulation of either CD4 or CD8 accompanies positive selection and precedes the emergence of mature cells with distinct effector functions.

[1]Department of Microbiology and Immunology, University of California at San Francisco, San Francisco, CA 94143-0414, USA
[2]Molecular Pathogenesis Program and The Howard Hughes Medical Institute, The Skirball Institute of Biomolecular Medicine, New York University Medical Center, 540 First Avenue, Second Floor, New York, NY 10016, USA

CD4 and CD8 bind to MHC class II and class I proteins, respectively, and they also both interact with the cytoplasmic tyrosine kinase p56$^{lck}$ (reviewed elsewhere in this volume). During T cell responses, CD4 and CD8 function as coreceptor molecules and are recruited to T cell receptors that are actively engaging antigen/ MHC complexes (JANEWAY 1992). As argued by KÖNIG et al. (this volume), the coreceptors are likely to provide crucial stability to weak affinity T cell receptor (TCR)-MHC interactions, which is essential for antigen-specific signal transduction. Recent experiments using transgenic and gene inactivation techniques in mice have provided insight into the roles of CD4 and CD8 in the selection of T cells with distinct phenotypes (ROTHENBERG 1994). Results of some of these studies, particularly those on the function of CD4, suggest that the coreceptors are responsible for the tight coupling that is maintained between the MHC class specificity of T lymphocytes and their effector functions. In the thymus, the extinction of CD4 and CD8 expression during positive selection appears to be tightly linked to programs of cytotoxic and helper cell differentiation, respectively. The coreceptors therefore appear to have evolved to provide the fine tuning required for efficient lineage commitment of T lymphocytes; this ensures that T cells which interact with antigen originating via a biosynthetic route within a target cell have cytotoxic function, while T cells that interact with antigen taken up by endocytic routes into antigen presenting cells have specialized helper functions.

## 2 Thymocyte Ontogeny

The development of αβ T cells in the thymus can be conveniently subdivided into three sequential stages (Fig. 1).

### 2.1 The Double Negative Cells

The first stage ($CD4^-CD8^-$; double negative; DN) is the most immature and includes progenitor cells which have migrated from the bone marrow to seed the thymic cortex. It also includes the direct descendants of these cells, some of which have activated the RAG-1 and RAG-2 genes (TURKA et al. 1991) and undergone rearrangement at the TCR β and TCR-γ loci. Irreversible commitment to the T cell developmental program may occur prior to entry into the thymus, but at least some of the early entrants retain the ability to contribute to other lineages including NK cells (SPITS 1994). Several discrete substages in the early differentiation process have been identified (GODFREY et al. 1994). For example, cells that lack expression of the interleukin-2 (IL-2) receptor α-chain and CD44 are derived from precursor cells that coexpress both of these molecules. Exit of TCR-αβ lineage cells out of the DN stage occurs following productive rearrangement at the TCR-β locus (MOMBAERTS et al. 1992, 1994), which results in surface expression of a β-chain associated with an invariant chain, gp33 pre-Tα (GROETTRUP et al. 1993).

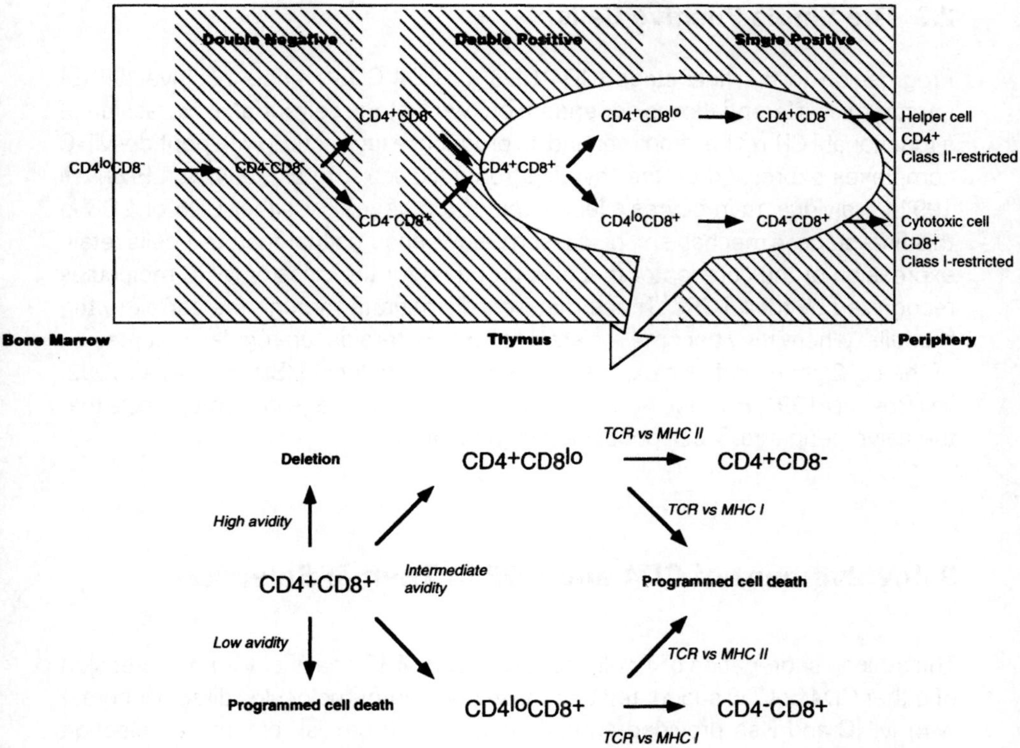


**Fig. 1.** Stages in the development of αβ T cells. The expanded section of the figure shows the formation of the two lineages of T cells within the thymus. The term "avidity" is defined in the text: high avidity interactions (involving strong coreceptor/T cell receptor (TCR) interactions with peptide/MHC) would lead to thymocyte deletion, whereas weak avidity interactions would not provide the necessary signals for developmental progression. $CD4^+CD8^+$ thymocytes which may express TCRs of intermediate avidity will also not complete development if they inappropriately down-regulate the coreceptor that would otherwise match the specificity of their TCRs. Adapted from BENOIST and MATHIS 1994)

## 2.2 The Double Positive Cells

Expression of the immature pre-T cell receptor (i.e., TCRβ+gp33) results in a signal that permits progression of DN cells to the DP stage, in which both CD4 and CD8 genes are transcriptionally active and DNA rearrangement occurs at the TCR-α locus (MOMBAERTS et al. 1992). The DP ($CD4^+CD8^+$) thymocytes are subjected to a selection process that shapes the T cell repertoire (VON BOEHMER 1994): the few cells that have useful TCRs, able to interact with host MHC molecules and bound peptides, are positively selected to complete maturation; those cells with nonproductive rearrangements of their TCR-α genes or with TCRs that do not interact with host MHC undergo apoptosis; and potentially self-reactive thymocytes, with receptors that react too vigorously with peptide/MHC complexes, are eliminated through activation of an apoptotic pathway. This selection process gives rise to mature T cells that can respond efficiently to foreign peptides bound to host MHC molecules.

### 2.3 The Single Positive Cells

Progression to the final stage ($CD4^+CD8^-$ or $CD4^-CD8^+$; single positive; SP) of intrathymic differentiation is dependent on two factors: cell surface expression of a functional TCR-αβ heterodimer and its productive interaction with peptide/MHC complexes expressed on the thymic stroma (HOGQUIST et al. 1994; VON BOEHMER 1994). Individual thymocytes silence one of the genes for either CD4 or CD8 in this final stage; a mechanism, discussed below, ensures that mature T cells retain expression of the coreceptor that is appropriate for the class of MHC molecules recognized by their TCRs. These mature cells migrate through the cortex to the medulla, where they become resistant to corticosteroids, up-regulate expression of the *bcl-2* gene, and lose expression of the HSA molecule (BENDELAC et al. 1992; VON BOEHMER 1994; ROTHENBERG 1994). It is from this final stage in thymopoiesis that the naive peripheral T cell population is generated.

## 3 Involvement of CD4 and CD8 in Thymic Selection

The striking segregation of T cell phenotype and MHC specificity with expression of either CD4 or CD8 suggested that these coreceptor molecules directly interact with MHC and also provided an early hint of their crucial role in the selection process (SWAIN 1983). Subsequent studies clearly showed that CD4 and CD8 bind to nonpolymorphic membrane-proximal regions of MHC class II and class I, respectively (see Chap. 2). Disruption of this binding in vivo, using monoclonal antibodies against CD4 or CD8 and gene ablation by homologous recombination, have confirmed that the coreceptor molecules have important roles in both the positive and negative selection of thymocytes. For example, neonatal injection of noncytolytic anti-CD4 and anti-CD8 monoclonal antibodies blocked the development of CD4+ and CD8+ cells, respectively (RAMSDELL and FOWLKES 1989; ZUÑIGA-PFLÜCKER et al. 1990). The deletion of autoreactive T cells was also shown to be blocked by anti-CD4 monoclonal antibodies (FOWLKES et al. 1988; MACDONALD et al. 1988). In mice deficient for CD4 or CD8 expression, the majority of cells of each respective lineage failed to develop (FÚNG-LEUNG et al. 1991; RAHEMTULLA et al. 1991; KILLEEN et al. 1993). Negative selection of some, but not all, self-reactive T cells was also hindered in these mutant mice (WALLACE et al. 1992; FUNG-LEUNG et al. 1993).

To be effective in signal transduction during T cell development or activation, the binding of CD4 or CD8 to MHC must coincide with TCR binding to the same MHC molecule (SALTER et al. 1990; CONNOLLY et al. 1990). This requirement was demonstrated most clearly using transgenic mice expressing either wild-type MHC class I molecules or variants that could not interact with murine CD8 (ALDRICH et al. 1991; INGOLD et al. 1991; KILLEEN et al. 1992; SCHONRICH et al. 1993). Interaction was abrogated by making point mutations in an exposed negatively charged loop of the α3-domain or by substituting the human HLA α3-domain for that in mouse class I. The mutant class I molecules were capable of presenting antigenic peptides efficiently to peripheral T cells, yet they did not allow for positive selection and, in some cases, for negative selection of thymocytes. These findings were consistent with the observation that CD8 lineage T cells fail to develop in CD8-null mice (FUNG-LEUNG et al. 1991). Importantly, selection in the mutant class I transgenic mice was disrupted despite coexpression of endogenous wild-type MHC class I proteins together with the non-CD8-binding versions. Therefore, these experiments demonstrated a requirement in thymic selection for coengagement of the same MHC complex by both the TCR and the CD8 molecule. This confirmed the role of CD8 as a coreceptor that makes an essential contribution to TCR recognition of peptide/MHC class I.

Analogous transgenic experiments have yet to be performed to formally demonstrate a coreceptor role for CD4 in its interaction with MHC class II during T cell development. However, numerous in vitro studies strongly suggest that CD4 has a role analogous to that of CD8 during T cell development and activation (reviewed in Chaps. 2, 3). For example, independent ligation of CD4 by antibody (BANK and CHESS 1985; TITE et al. 1986; WASSMER et al. 1985; BEKOFF et al. 1985; NEWELL et al. 1990) inhibits activation of T cells upon separate ligation of the TCR, whereas T cell activation can proceed when these molecules are coaggregated. This suggests that, for effective signaling in vivo, there must be corecognition of MHC class II by the TCR and CD4. The role of coengagement of MHC by TCRs with either coreceptor is discussed in detail below, in the context of an avidity model for thymocyte selection.

## 4 Mechanism of Coreceptor Function During T Cell Development

A series of recent studies on the role of the CD4 and CD8 molecules in developmental signal transduction have provided evidence that these molecules function by contributing to the overall stability of the TCR-MHC interaction (LEE et al. 1992; ROBEY et al. 1992; KILLEEN and LITTMAN 1993; CHAN et al. 1993a). In addition, down-regulation of either CD4 or CD8 during positive selection is coupled to activation of a specific functional program of differentiation (CORBELLA et al. 1994). Developing thymocytes that demonstrate the required binding characteristics for MHC/peptide complexes and complete positive selection are hence likely to have functions appropriate for their MHC specificities. The results of the studies are as follows:

(1) The Lck molecule associated with CD4 has a function that is independent of its kinase activity, but dependent on its SH2 and SH3 domains, which are involved in protein-protein interactions (XU and LITTMAN 1993). Stimulation of the TCR complex, resulting in recruitment of CD4 to the complex, requires the association of Lck with CD4 (COLLINS et al. 1992). Together, these results

suggest that CD4-associated Lck functions to recruit CD4 to partially activated TCRs.

(2) The Lck-binding cytoplasmic domain of CD4 is not absolutely necessary for CD4 function in positive selection (KILLEEN and LITTMAN 1993); in its absence, only high concentrations of CD4 will permit development of T helper lineage cells. Similarly, overexpression of a CD8 α molecule that cannot bind to Lck also rescues development of cytotoxic lineage T cells (CHAN et al. 1993a). Together, results (1) and (2) are consistent with a role for Lck in the recruitment of CD4 to the TCR complex, where it can contribute to stabilizing the interaction between the TCR and peptide/MHC class II. The transgenic experiments described in KILLEEN and LITTMAN (1993) and CHAN et al. (1993a) suggest that a signaling role for coreceptor-associated Lck can be dispensable for positive selection.

(3) In the absence of CD4, a considerable fraction (~15%) of the T helper repertoire, specific for MHC class II, can complete development and leave the thymus (LOCKSLEY et al. 1993; RAHEMTULLA et al. 1994). These cells are programmed to transcribe CD4, despite the disruption of the coding sequence by the inactivating mutation (KILLEEN and LITTMAN, in preparation). Significantly, the majority of CD4 lineage cells cannot overcome the absence of CD4 and do not exit from the thymus. The basis for this differential dependence on CD4 remains to be resolved, but one possible explanation is that cells which can complete development in the absence of CD4 are distinguished by expression of TCRs that have high affinity for peptide/MHC. Such high affinity receptors would transmit a positive selection signal without the need for the stabilizing contribution of the CD4 coreceptor. In contrast, the majority of MHC class II-restricted receptors that are normally expressed by $CD4^+$ thymocytes would be of weaker affinity and therefore would require CD4 coreceptor activity for positive selection.

(4) Overexpression of CD4 causes clonal deletion of thymocytes bearing a transgenic TCR that would normally direct positive selection of cells expressing wild-type levels of CD4 (DAVIS and LITTMAN, submitted). Similar results have been reported upon overexpression of CD8 in transgenic mice (LEE et al. 1992; ROBEY et al. 1992). From these studies and those summarized in (3), it is clear that the level of coreceptor expression influences the outcome of thymocyte selection. Furthermore, it is likely that the fate of a developing thymocyte is determined by the potency of TCR stimulation which in turn is dependent on the affinity of the TCR for peptide/MHC complexes, the concentration of such complexes on the thymic stroma, and the level of TCR and coreceptor expression on individual thymocytes. If the term avidity is used to quantitate the aggregate influence of all of these variables, then thymocyte positive selection occurs only when this avidity falls between two thresholds: a lower threshold beneath which cells undergo programmed cell death due to lack of stimulation; and an upper threshold above which autoreactive cells that respond too vigorously to thymic peptide/MHC are deleted by induction of an apoptotic pathway (Fig. 1).

(5) Ectopic expression of either CD4 or CD8 results in the appearance of mature thymocytes and T cells that have mismatched MHC specificities of their TCR and endogenous coreceptor (Davis et al. 1993; Itano et al. 1994; Baron et al. 1994; Chan et al. 1994; Robey et al. 1994); these results indicate that a major component of the lineage decision to become a CD4 or CD8 SP cell is achieved stochastically, and is followed by selection of cells with the appropriate match of TCR and coreceptor specificity for MHC (Fig. 1).

(6) Rescue of cells with mismatched TCR and coreceptor (e.g., class I-specific CD4 SP cells) results in a mismatch of the MHC specificity (class I) and the function of the T cell (helper function), indicating that there is coupling between down-regulation of CD4 or CD8 and commitment to a functional lineage (Davis et al. 1993; Corbella et al. 1994; Chan et al. 1994; Baron et al. 1994; Robey et al. 1994). This indicates that function is determined prior to completion of positive selection and is independent of the TCR interaction with a particular class of MHC.

Taken together, it can be argued that the fate of a developing thymocyte is most influenced by avidity as defined in (4). The initial event that rescues DP thymocytes from programmed cell death is an interaction of sufficiently high avidity between TCR/coreceptor and MHC/peptides (Chan et al. 1994; Killeen and Littman, in preparation). This first trigger of positive selection results in the stochastic down-regulation of one of the coreceptors and coupled commitment to one of the functional lineages. Loss of the appropriate coreceptor results in a sub threshold avidity and death of the cell; continued expression of the correct coreceptor retains the requisite avidity and results in full development of the T cell. If this avidity is too high, the cell would activate an apoptotic process, thus aborting positive selection. This model does not rule out the possibility that the process is rendered more efficient by specific instructional signals mediated via CD4 or CD8 (see below).

## 5 Regulation of CD4 Gene Expression

During intrathymic development, expression of the CD4 and CD8 glycoproteins is regulated in an ordered fashion. Both molecules are coordinately up-regulated on DP cells and are subsequently shut off on mature T cells in the course of positive selection (Fig. 1). Some studies have suggested that the regulation of coreceptor gene expression is at the level of mRNA stability (Paillard et al. 1990; Takahama and Singer 1992). For example, activation of DP thymocytes with phorbol esters or anti-TCR antibodies results in CD4 and CD8 mRNA down-regulation without affecting the rate of CD4 transcription (Takahama and Singer 1992). If this process is similar to thymocyte activation during positive selection, it would suggest that CD4 expression is regulated at the post-transcriptional level, at least during the early transition from DP to SP thymocytes. A more recent study has shown that

CD4 expression on mature T cells is regulated at the transcriptional level (Siu et al. 1994). It is theoretically possible that, early in development, CD4 down-regulation is post-transcriptional while later (in peripheral T cells) it occurs through a shut-off in transcription. However, thymocytes down-regulating CD4 following anti-TCR treatment are unlikely to be equivalent to cells undergoing positive selection, and other results, described below, are difficult to reconcile with a post-transcriptional mechanism.

Recent studies, using both transfection assays and analyses of transgenic mice, provide strong evidence that CD4 expression is regulated at the transcriptional level (Sawada and Littman 1991, 1993; Sawada et al. 1994; Siu et al. 1994). These experiments have identified elements that are involved in both positive and negative regulation of CD4 gene transcription. The initial study identified eight DNase I hypersensitive sites within 80 kb of the murine CD4 locus (Fig. 2). Only one of these sites, located 13 kb upstream of the transcription initiation site, could function as an enhancer in transient transfection assays when combined with the CD4 promoter region (Sawada and Littman 1991). This enhancer was specifically active in T cells and inactive in B lymphocytes and fibroblasts. However, the enhancer functioned in T cells regardless of whether or not they expressed CD4, suggesting that additional elements must be involved in T cell subset-specific gene expression.

Dissection of the enhancer region narrowed it down to a 339 bp fragment containing one binding site for the HMG family member LEF-1 (Travis et al. 1991) and three separate E-box sites for binding of bHLH family members (Weintraub et al. 1991). While binding of LEF-1 was largely dispensable for activity in transfection assays, binding of a T cell-specific factor to one of the E-boxes (Fig. 2) was found to be required. The protein complex binding to this E-box was shown to include a heterodimer (or hetero-oligomer) of the transcription factors HEB and E12/E47 (Sawada and Littman 1993). The T cell specificity of this nucleoprotein complex can probably be explained by the preferential expression of HEB in T cells (Hu et al. 1992) even though E12/E47 is ubiquitously expressed (Murre et al. 1989).

The tissue specificity of the CD4 promoter has been the subject of some controversy. Two groups have reported that the TATA-less promoter region displays T cell subset-specific transcriptional activity (Siu et al. 1992; Salmon et al. 1993). One of these groups has shown that binding sites for Ets family members are important for promoter activity and that $CD4^+$ cells specifically express an Ets-related protein that binds to one of these sites (Salmon et al. 1993). The other group has shown that a Myb binding site in the CD4 promoter is required for full promoter function (Siu et al. 1992). The finding of subset specificity in these studies is difficult to reconcile with the observation that a combination of CD4 enhancer and promoter directs reporter gene expression in both CD4 SP and CD8 SP cell lines (Sawada and Littman 1991). In addition, when combined with the SV40 enhancer, the CD4 promoter was active in non-T cells (Sawada and Littman 1991). More recent studies in transgenic systems further indicate that, in the presence of the CD4 enhancer, the CD4 promoter is not

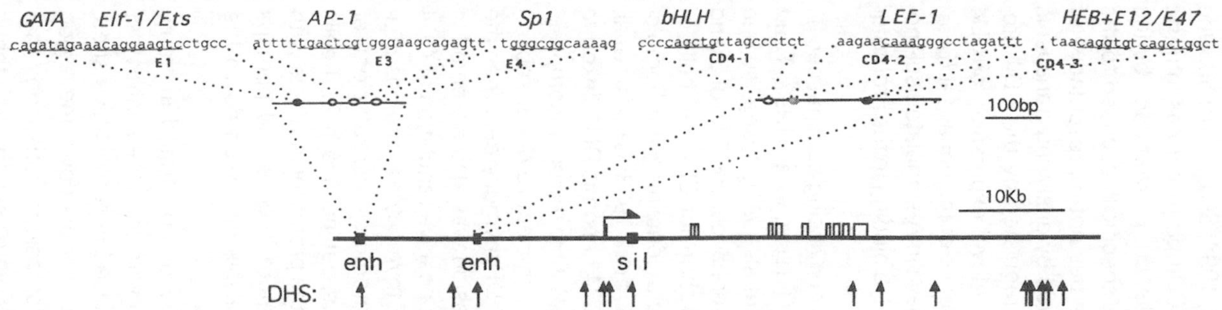


**Fig. 2.** Regulatory elements within the murine CD4 locus. DNAse I hypersensitivity sites (DHS) that map within the murine CD4 gene are shown as *vertical arrows* (from SAWADA and LITTMAN 1991; SANDS and NIKOLIC-ZUGIC 1992; SIU et al. 1994). Exons are shown as *open boxes* and the transcriptional start site is indicated by a *horizontal arrow*. Proximal and distal enhancers (enh) and the silencer element (sil) are shown as filled boxes (SAWADA and LITTMAN 1991, 1993; WURSTER et al. 1994). The expanded section shows some of the factor-binding sites within the two enhancers: *open symbols* denote motifs that can be mutated with negligible effects on transcriptional enhancer activity in vitro; *solid symbols* (for E1 and CD4–3) denote motifs that are essential for enhancer function; mutations which disable the Lef-1 binding site (CD4-2; *shaded symbol*) of the proximal enhancer have slight effects on enhancer activity. Sequences matching the consensus for the indicated transcription factors are *underlined*. The binding of Elf-1, Lef-1 and HEB+E12/E47 to E1, CD4–2 and CD4–3, respectively, has been confirmed by electrophoretic mobility shift analysis using specific antisera or by southwestern blotting. The proximal enhancer has been shown to be required for efficient expression of CD4 transgenes in mice (KILLEEN et al. 1993; BLUM et al. 1993; GILLESPIE et al. 1993; HANNA et al. 1994). Transgenes that include the silencer element are active in mature cells of the CD4 lineage but not of the CD8 lineage (SAWADA et al. 1994; SIU et al. 1994)

sufficient for subset-specific expression (SAWADA et al. 1994; SIU et al. 1994). One possible explanation for this discrepancy is that the CD4 promoter has intrinsic preference for expression in the CD4 subset of T cells, but is overridden by the presence of the CD4 enhancer. Alternatively, subset specificity ascribed to the CD4 promoter may be due to cell lines having properties not representative of mature T cell subsets. This issue needs to be resolved by analyzing transgenic mice prepared with an enhancerless reporter gene regulated by the CD4 promoter. Such a study should be feasible because low level expression of CD4 transgenes has been occasionally observed in the absence of the enhancer (BLUM et al. 1993; SAWADA et al. 1994), particularly in concanavalin A-stimulated splenocytes (HANNA et al. 1994). The significance of Ets and Myb binding to the promoter region remains to be determined.

A series of transgenic studies has demonstrated convincing subset-specific expression of CD4 constructs (KILLEEN et al. 1993; GILLESPIE et al. 1993; BLUM et al. 1993; HANNA et al. 1994; SAWADA et al. 1994; SIU et al. 1994). A large human CD4 transgenic clone containing the upstream murine enhancer and 3 kb of 5' and 8 kb of 3' flanking sequence was expressed in CD4, but not CD8, SP cells of transgenic mice (KILLEEN et al. 1993). Other constructs containing the homologous human enhancer were similarly expressed (BLUM et al. 1993; GILLESPIE et al. 1993). Using fusion constructs consisting of human CD4 genomic DNA linked to 3' cDNA sequence, HANNA et al. (1994) showed that a 12.6 kb region encompassing 2.6 kb 5' flanking sequence through exon 3 was sufficient to direct subset-specific expression in mature CD4 SP T cells, but not in DP thymocytes. When the murine enhancer was linked to this construct, expression was also observed in the immature thymocytes. This result suggested that positive regulation of CD4 expression may differ between mature T cells and DP thymocytes. The observation was also reminiscent of the finding that the CD4 promoter was sufficient for subset-specific expression in transfected SP T cell lines, but inactive in DP T cells (SIU et al. 1992). It is thus possible that a region within the promoter interacts with transcription factors that function exclusively in mature T cells and perhaps preferentially in CD4 SP T cells. In addition, this analysis narrowed down the region required for subset-specific expression of CD4.

Reliable subset-specific expression could not be demonstrated in transfection studies when the CD4 enhancer and promoter were combined with other sequences in the CD4 gene that corresponded to DNase I hypersensitive sites. Transgenic studies were therefore employed to further narrow down the region containing elements required for CD4 SP-specific expression. Transgenes that incorporated either human CD2 or HLA-B7 as reporters were used to show that a region within intron 1 of the murine CD4 gene conferred subset-specific expression when combined with the CD4 enhancer and promoter (SAWADA et al. 1994; SIU et al. 1994). By using a minigene consisting of the 339 bp enhancer, 0.5 kb promoter, noncoding exon 1, pieces of intron 1, and exon 2 (in which the hCD2 reporter gene was introduced), it was possible to demonstrate that a 0.4 kb fragment of intron 1 was sufficient to shut off expression of the reporter gene in CD8 SP cells (SAWADA et al. 1994). This intronic element corresponds to a previously described DNase I hypersensitive site (SAWADA and LITTMAN 1991; SANDS and NIKOLIC-ZUGIC 1992). In its absence, hCD2 was expressed not only on CD8 SP T cells, but also in DN thymocytes. This indicated that the CD4 enhancer is active in early thymocyte development but is normally countered by the silencing effect of the intronic sequence (SAWADA et al. 1994). Interestingly, mouse CD4 is expressed on very immature prothymocytes (WU et al. 1991) and precursor cells in the bone marrow (FREDERICKSON and BASCH 1989). Silencing must therefore occur during at least two stages of T cell development, in early DN cells and in mature CD8 SP cells. A mechanism for inactivation of the silencer must then be employed during the DN to DP transition.

The combination of the minimal enhancer, promoter, and silencer directed expression of the hCD2 reporter during thymocyte development in a fashion that was indistinguishable from that observed for endogenous CD4 (SAWADA et al. 1994). It is, therefore, highly unlikely that post transcriptional mechanisms have a significant role in the activation and down-regulation of CD4 expression in the thymus.

When a combination of the T cell-specific enhancer and a more distal enhancer element, located about 25 kb 5' of the transcription initiation site (WURSTER et al. 1994), was used in the transgenic constructs, expression was observed not only in T cells, but also in B cells and macrophages (SIU et al. 1994). If the intronic element was included, expression was limited to T lymphocytes, indicating that the silencer can function in non-T lineage cells and suggesting that there may be potential for expression of CD4 in cells of the B and macrophage lineages. Deletion of the silencer by homologous recombination in the mouse genome will be required to determine if this potential can be realized when the positive-acting elements are in their normal genomic configuration.

The intronic element was shown to be a bona fide silencer by a number of established criteria. Its activity was independent of orientation (SAWADA et al. 1994); it functioned when placed either upstream or downstream of the transcribed region (SAWADA et al. 1994; SIU et al. 1994); and it was active in combination with a heterologous enhancer and promoter (obtained from the CD3δ gene; SAWADA et al. 1994). The mechanism through which silencing is achieved is not yet understood. It may involve locus inactivation through changes in the structure of adjacent chromatin; or it may be due to direct interaction of a silencing factor with the components involved in initiating transcription of the CD4 gene.

Transcriptional silencing by the CD4 intronic element has been reproduced in tissue culture, but only when multiple copies of the sequence were included (SAWADA et al. 1994). Thus, in the absence of the 0.4 kb silencer, the CD4 enhancer and promoter directed expression of the CAT reporter in both $CD4^+ CD8^+$ (e.g., 1010) and $CD4^- CD8^+$ (1200M) T cell lines. When three copies of the silencer sequence were included, there was specific repression observed only in the CD8 SP 1200M cells. This finding may open the way for the biochemical isolation of specific factors involved in silencing.

In human and rat, but not in mouse, the CD4 glycoprotein is expressed on both macrophages and on other hematopoietic lineage cells, such as eosinophils

(Crocker et al. 1987; Kazazi et al. 1989; Lucey et al. 1989). The absence of CD4 expression in mouse monocytes and macrophages is due to a species difference in *cis*-acting sequences of the CD4 genes. Human CD4 transgenes were thus expressed in murine macrophages, indicating that the requisite *trans*-acting factors are present in mouse (Hanna et al. 1994; Killeen et al., unpublished results). Expression in macrophages could be demonstrated with a transgene containing only 2.6 kb upstream of the human CD4 transcription start site plus exons 1–3 and the intervening introns (Hanna et al. 1994). The T cell-specific upstream CD4 enhancer was not required for macrophage-specific expression. Elucidation of the mechanism of macrophage-specific CD4 expression may facilitate approaches aimed at containing the pathogenic effects of HIV, which infects both T cells and macrophages.

## 6 Role of CD4 in Thymocyte Lineage Specification

The mechanism of lineage choice during which DP thymocytes differentiate into either CD4 or CD8 SP cells has been the subject of intense investigation and considerable controversy. A number of changes occur upon delivery of a positive selection signal: the level of surface TCR/CD3 is up-regulated considerably through a poorly understood post translational mechanism (Kearse et al. 1994); bcl-2 expression is up-regulated (Moore et al. 1994; Veis et al. 1993; Gratiot-Deans et al. 1993); the activation marker CD69 is transiently expressed (Swat et al. 1993; Bendelac et al. 1992) and there are many other cell surface changes including the down-regulation of the heat stable antigen (HSA). At this time, CD4 or CD8 levels are down-regulated through a transcriptional inactivation mechanism, as described above, and cells commit to either the CD4 or CD8 lineages. Two possible mechanisms could direct this lineage commitment: either the down-regulation of co-receptors is strictly dictated by MHC specificity of the TCR; or, alternatively, it may be independent of TCR specificity and thus may occur stochastically (Fig. 1).

Several early studies indirectly suggested that there must be an instructional signal that directs the cell to either the CD4 or CD8 lineage (Robey et al. 1991; Borgulya et al. 1991; Seong et al. 1992). This would be most readily explained by distinct signals initiated upon coengagement of CD4 or CD8 and the TCR with MHC class II or class I on thymic epithelium. Subsequent experiments, including the finding, discussed above, that at least a component of helper T cell development is completely independent of CD4-mediated signaling, have provided support for the opposing scenario (Davis and Littman 1994). A consensus has now emerged that at least part of the mechanism of lineage commitment is due to a stochastic process.

This fundamental developmental problem has been studied in two ways. The first was to demonstrate that, in MHC-null mice, thymocytes could be found that paradoxically had a phenotype of cells that interact with the missing class of MHC molecules (CHAN et al. 1993b; VAN MEERWIJK and GERMAIN 1993). Thus, in MHC class II negative mice, in which no mature $CD4^+$ T cells are observed, thymocytes that had begun to down-regulate CD8 but continued to express high levels of CD4 could be detected; these cells appeared to be intermediates in positive selection, as they had up-regulated the level of TCR (to an intermediate level) and of CD69 but had not yet down-regulated HSA or acquired resistance to corticosteroids (CHAN et al. 1993b). Despite having the phenotype of cells that recognize MHC class II, these cells were specific for class I. Conversely, in class I ($\beta_2$m)-null mice, a similar population that had down-regulated CD4 was observed (CHAN et al. 1993b; VAN MEERWIJK and GERMAIN 1993). Mice expressing neither MHC molecule lacked the intermediate population (CHAN et al. 1993b; CRUMP et al. 1993). This suggested that engagement of the TCR by MHC would trigger the thymocyte to commit to differentiation along either the CD4 or CD8 pathway. In the absence of the MHC molecules required for further differentiation and completion of positive selection, the committed population would be arrested and would die.

This interpretation is supported by results of a second series of experiments, in which ectopic expression of coreceptors was shown to rescue the intermediate population of stochastically generated committed thymocytes that had mismatched coreceptor and TCR specificity. Thus, in mice lacking class I-specific T cells (due to absence of $\beta_2$m and/or expression of a class II-specific transgenic TCR), CD8 lineage cells with class II-specific TCRs were rescued upon expression of a CD4 transgene (DAVIS et al. 1993; BARON et al. 1994). The rescued cells, despite being class II-specific, had a cytotoxic T cell phenotype, indicating that there is coupling between coreceptor down-regulation and a functional program of differentiation (DAVIS et al. 1993; BARON et al. 1994; DAVIS and LITTMAN, manuscript in preparation).

Several other groups have now confirmed these results in CD4 lineage commitment as well. Initial studies failed to demonstrate rescue upon expression of CD8 in mice bearing the transgenic TCR specific for H-Y/H-$2D^b$ (ROBEY et al. 1991; BORGULYA et al. 1991). This result was initially interpreted as providing support for an instructional mechanism, as described above. Subsequent experiments have shown that CD4 lineage cells bearing mismatched class I-specific TCRs can in fact survive upon expression of transgenic CD8$\alpha\beta$ molecules (ITANO et al. 1994; CHAN et al. 1994; ROBEY et al. 1994). Moreover, these cells have a helper phenotype (CORBELLA et al. 1994; ROBEY et al. 1994). In MHC class II-null mice, expression of CD8 thus not only rescued appearance of $CD4^+$ cells, but also resulted in the appearance of germinal centers, which are not observed in the absence of T helper cells (CHAN et al. 1994). Taken together, the results are now consistent in demonstrating that there is a stochastic component to lineage commitment.

Rescue with forced expression of coreceptor is relatively inefficient, suggesting that there may yet be some instructional signaling involved in this complex process (DAVIS and LITTMAN 1994; ROBEY et al. 1994). It is possible that, in the absence of signaling through either coreceptor, there would be a baseline stochastic process resulting in commitment to the CD4/$T_h$ or CD8/$T_c$ lineage. This may function through a series of positive and negative feedback loops, ensuring that only one of the two pathways of differentiation prevails. A discriminatory signal through either CD4 or CD8, resulting from their cross-linking to the TCR, would preferentially affect one of the two pathways, greatly increasing the probability that it would prevail. To date, there has not been any direct evidence for such instructional signaling.

Based on the results described above, we suggest that the role of CD4 and CD8 is to insure that the majority of T cells have a function that is appropriate for the type of MHC molecule that they recognize (DAVIS and LITTMAN 1994). Since coreceptor expression is linked to the functional phenotype of the T cell, there would be few cells whose MHC specificity does not match the function (these cells would, presumably, have high affinity TCR-MHC interactions, and would therefore not require the coreceptor to complete positive selection). Superimposition of a degree of instructional signaling may further reduce the number of such mismatched T cells.

## 7 Summary

The data reviewed in this chapter suggest that the primary developmental function of the CD4 and CD8 coreceptors is to improve the efficacy by which a thymocyte recognizes peptide/MHC. During positive selection, DP thymocytes down-regulate expression of either CD4 or CD8 in response to signals that originate from the TCR/coreceptor complex. Experiments with transgenic and MHC-null mice have shown that coreceptor down-regulation and lineage commitment can occur stochastically in a manner that is independent of TCR specificity for MHC. Nevertheless, the positive selection of a given thymocyte is contingent on sustained expression of the coreceptor that is appropriate for the MHC specificity of its TCR. In most cases, loss of the required coreceptor blocks developmental progression and results in thymocyte apoptosis.

CD4 expression is controlled by both positive and negative regulatory sequences embedded in the CD4 gene and it is likely that similar sequences regulate the CD8 gene. The down-regulation of coreceptor expression is coupled to a functional commitment which ensures that mature $CD4^+$ T cells have a helper phenotype and $CD8^+$ T cells have a cytotoxic phenotype. The molecular basis for this coupling and the identity of the switching mechanism which governs coreceptor regulation remain to be determined.

*Acknowledgments.* DRL is supported by the Howard Hughes Medical Institute. NK is supported by a Special Fellowship from the Leukemia Society. The UCSF transgenic mouse facility was supported in part by a grant from the Markey Foundation.

# References


Aldrich CJ, Hammer RE, Jones-Youngblood S, Koszinowski U, Hood L, Stroynowski I, Forman J (1991) Negative and positive selection of antigen-specific cytotoxic T lymphocytes affected by the alpha 3 domain of MHC I molecules. Nature 352: 718–721

Bank I, Chess L (1985) Perturbation of the T4 molecule transmits a negative signal to T cells. J Exp Med 162: 1294–1303

Baron A, Hafen K, von Boehmer H (1994) A human CD4 transgene rescues CD4-CD8+ cells in beta 2-microglobulin-deficient mice. Eur J Immunol 24: 1933–1936

Bekoff M, Kakiuchi T, Grey HM (1985) Accessory cell function in the Con A response: role of Ia-positive and Ia-negative accessory cells. J Immunol 134: 1337–1342

Bendelac A, Matzinger P, Seder RA, Paul WE, Schwartz RH (1992) Activation events during thymic selection. J Exp Med 175: 731–742

Benoist C, Mathis D (1994) T cell receptors-the end of a frustrating search. Curr Biol 4: 821–823

Blum MD, Wong GT, Higgins KM, Sunshine MJ, Lacy E (1993) Reconstitution of the subclass-specific expression of CD4 in thymocytes and peripheral T cells of transgenic mice: identification of a human CD4 enhancer. J Exp Med 177: 1343–1358

Borgulya P, Kishi H, Muller U, Kirberg J, von Boehmer H (1991) Development of the CD4 and CD8 lineage T cells: instruction versus selection. EMBO J 10: 913–918

Chan IT, Limmer A, Louie MC, Bullock ED, Fung-Leung W-P, Mak TW, Loh DY (1993a) Thymic selection of cytotoxic T cells independent of CD8α-Lck association. Science 261: 1581–1584

Chan SH, Cosgrove D, Waltzinger C, Benoist C, Mathis D (1993b) Another view of the selective model of thymocyte selection. Cell 73: 225–236

Chan SH, Waltzinger C, Baron A, Benoist C, Mathis D (1994) Role of coreceptors in positive selection and lineage commitment. EMBO J 13: 4482–4489

Collins TL, Uniyal S, Shin J, Strominger JL, Mittler RS, Burakoff SJ (1992) $p56^{lck}$ association with CD4 is required for the interaction between CD4 and the TCR/CD3 complex and for optimal antigen stimulation. J Immunol 148: 2159–2162

Connolly JM, Hansen TH, Ingold AL, Potter TA (1990) Recognition by CD8 on cytotoxic T lymphocytes is ablated by several substitutions in the class I alpha 3 domain: CD8 and the T-cell receptor recognize the same class I molecule. Proc Natl Acad Sci USA 87: 2137–2141

Corbella P, Moskophidis D, Spanopoulou E, Mamalaki C, Tolaini M, Itano A, Lans D, Baltimore D, Robey E, Kioussis D (1994) Functional commitment to helper T cell lineage precedes positive selection and is independent of T cell receptor MHC specificity. Immunity 1: 269–276

Crocker PR, Jefferies WA, Clark SJ, Chung LP, Gordon S (1987) Species heterogeneity in macrophage expression of the CD4 antigen. J Exp Med 166: 613–618

Crump AL, Grusby MJ, Glimcher LH, Cantor H (1993) Thymocyte development in major histocompatibility complex-deficient mice: evidence for stochastic commitment to the CD4 and CD8 lineages. Proc Natl Acad Sci USA 90: 10739–10743

Davis CB, Littman DR (1994) Thymocyte lineage commitment: is it instructed or stochastic? Curr Opin Immunol 6: 266–272

Davis CB, Killeen N, Crooks MEC, Raulet D, Littman DR (1993) Evidence for a stochastic mechanism in the differentiation of mature subsets of T lymphocytes. Cell 73: 237–247

Fowlkes BJ, Schwartz RH, Pardoll DM (1988) Deletion of self-reactive thymocytes occurs at a $CD4^+8^+$ precursor stage. Nature 334: 620–623

Frederickson GG, Basch RS (1989) L3T4 antigen expression by hemopoietic precursor cells. J Exp Med 169: 1473–1478

Fung-Leung WP, Schilham MW, Rahemtulla A, Kundig TM, Vollenweider M, Potter J, van Ewijk W, Mak TW (1991) CD8 is needed for development of cytotoxic T cells but not helper T cells. Cell 65: 443–449

Fung-Leung WP, Wallace VA, Gray D, Sha WC, Pircher H, Teh HS, Loh DY, Mak TW (1993) CD8 is needed for positive selection but differentially required for negative selection of T cells during thymic ontogeny. Eur J Immunol 23: 212–216

Gillespie FP, Doros L, Vitale J, Blackwell C, Gosselin J, Snyder BW, Wadsworth SC (1993) Tissue-specific expression of human CD4 in transgenic mice. Mol Cell Biol 13: 2952–2958

Godfrey DI, Kennedy J, Mombaerts P, Tonegawa S, Zlotnik A (1994) Onset of TCR-beta gene rearrangement and role of TCR-beta expression during $CD3^-CD4^-CD8^-$ thymocyte differentiation. J Immunol 152: 4783–4792

Gratiot-Deans J, Ding L, Turka LA, Nunez G (1993) bcl-2 proto-oncogene expression during human T cell development. Evidence for biphasic regulation. J Immunol 151: 83–91

Groettrup M, Ungewiss K, Azogui O, Palacios R, Owen MJ, Hayday AC, von Boehmer H (1993) A novel disulfide-linked heterodimer on pre-T cells consists of the T cell receptor beta chain and a 33 kd glycoprotein. Cell 75: 283–294

Hanna Z, Simard C, Laperriere A, Jolicoeur P (1994) Specific expression of the human CD4 gene in mature $CD4^+$ $CD8^-$ and immature $CD4^+CD8^+$ T cells and in macrophages of transgenic mice. Mol Cell Biol 14: 1084–1094

Hogquist KA, Jameson SC, Bevan MJ (1994) The ligand for positive selection of T lymphocytes in the thymus. Curr Opin Immunol 6: 273–278

Hu JS, Olson EN, Kingston RE (1992) HEB, a helix-loop-helix protein related to E2A and ITF2 that can modulate the DNA-binding ability of myogenic regulatory factors. Mol Cell Biol 12: 1031–1042

Ingold AL, Landel C, Knall C, Evans GA, Potter TA (1991) Co-engagement of CD8 with the T cell receptor is required for negative selection. Nature 352: 721–723

Itano A, Kioussis D, Robey E (1994) Stochastic component to development of class I major histocompatibility complex-specific T cells. Proc Natl Acad Sci USA 91: 220–224

Janeway CA Jr (1992) The T cell receptor as a multicomponent signalling machine. Annu Rev Immunol 10: 645–674

Kazazi F, Mathijs JM, Foley P, Cunningham AL (1989) Variations in CD4 expression by human monocytes and macrophages and their relationships to infection with the human immunodeficiency virus. J Gen Virol 70: 2661–2672

Kearse KP, Roberts JL, Munitz TI, Wiest DL, Nakayama T, Singer A (1994) Developmental regulation of alpha-beta T cell antigen receptor expression results from differential stability of nascent TCR-alpha proteins within the endoplasmic reticulum of immature and mature T cells. EMBO J 13: 4504–4514

Killeen N, Littman DR (1993) Helper T cell development in the absence of CD4-$p56^{lck}$ association. Nature 364: 729–732

Killeen N, Moriarty A, Teh HS, Littman DR (1992) Requirement for CD8-MHC class I interaction in positive and negative selection of developing T cells. J Exp Med 176: 89–97

Killeen N, Sawada S, Littman DR (1993) Regulated expression of human CD4 rescues helper T cell development in mice lacking expression of endogenous CD4. EMBO J 12: 1547–1553

Kruisbeek AM (1993) Development of $\alpha\beta$ T cells. Curr Opin Immunol 5: 227–234

Lee NA, Loh DY, Lacy E (1992) CD8 surface levels alter the fate of $\alpha\beta$ TCR expressing thymocytes in transgenic mice. J Exp Med 175: 1013–1025

Locksley RM, Reiner SL, Hatam F, Littman DR, Killeen N (1993) Helper T cells without CD4: control of leishmaniasis in CD4-deficient mice. Science 261: 1448–1451

Lucey DR, Dorsky DI, Nicholson-Weller A, Weller PF (1989) Human eosinophils express CD4 protein and bind human immunodeficiency virus 1 gp120. J Exp Med 169: 327–332

MacDonald HR, Hengartner H, Pedrazzini T (1988) Intrathymic deletion of self-reactive cells prevented by neonatal anti-CD4 antibody treatment. Nature 335: 174

Mombaerts P, Clarke AR, Rudnicki MA, Iacomini J, Itohara S, Lafaille JJ, Wang L, Ichikawa Y, Jaenisch R, Hooper ML, Tonegawa S (1992) Mutations in T-cell antigen receptor genes alpha and beta block thymocyte development at different stages. Nature 360: 225–231

Mombaerts P, Anderson SJ, Perlmutter RM, Mak TW, Tonegawa S (1994) An activated lck transgene promotes thymocyte development in RAG-1 mutant mice. Immunity 1: 261–267

Moore NC, Anderson G, Williams GT, Owen JJ, Jenkinson EJ (1994) Developmental regulation of bcl-2 expression in the thymus. Immunology 81: 115–119

Murre C, McCaw PS, Baltimore D (1989) A new DNA binding and dimerization motif in immunoglobulin enhancer binding, daughterless, MyoD, and myc proteins. Cell 56: 777–783

Newell MK, Haughn LJ, Maroun CR, Julius MH (1990) Death of mature T cells by separate ligation of CD4 and the T cell receptor for antigen. Nature 347: 286–289

Paillard F, Sterkers G, Vaquero C (1990) Transcriptional and post-transcriptional regulation of TcR, CD4 and CD8 gene expression during activation of normal human T lymphocytes. EMBO J 9: 1867–1872

Rahemtulla A, Fung-Leung WP, Schilham MW, Kundig TM, Samhara SR, Narendran A, Arabian A, Wakeham A, Paige CJ, Zinkernagel RM, Miller RG, Mak TW (1991) Normal development and function of $CD8^+$ cells but markedly decreased helper cell activity in mice lacking CD4. Nature 353: 180–184

Rahemtulla A, Kundig TM, Narendran A, Bachmann MF, Julius M, Paige CJ, Ohashi PS, Zinkernagel RM, Mak TW (1994) Class II major histocompatibility complex-restricted T cell function in CD4-deficient mice. Eur J Immunol 24: 2213–2218

Ramsdell F, Fowlkes BJ (1989) Engagement of CD4 and CD8 accessory molecules is required for T cell maturation. J Immunol 143: 1467–1471

Robey E, Fowlkes BJ, Gordon JW, Kioussis D, von Boehmer H, Ramsdell F, Axel R (1991) Thymic selection in CD8 transgenic mice supports an instructive model for commitment to a CD4 or CD8 lineage. Cell 64: 99–107

Robey E, Itano A, Fanslow WC, Fowlkes BJ (1994) Constitutive CD8 expression allows inefficient maturation of CD4+ helper T cells in class II major histocompatibility complex mutant mice. J Exp Med 179: 1997–2004

Robey EA, Ramsdell F, Kioussis D, Sha WC, Loh DY, Axel R, Fowlkes BJ (1992) The level of CD8 expression can determine the outcome of thymic selection. Cell 69: 1089–1096

Rothenberg EV (1994) Signaling mechanisms in thymocyte selection. Curr Opin Immunol 6: 257–265

Salmon P, Giovane A, Wasylyk B, Klatzmann D (1993) Characterization of the human CD4 gene promoter: transcription from the CD4 gene core promoter is tissue-specific and is activated by Ets proteins. Proc Natl Acad Sci USA 90: 7739–7743

Salter RD, Benjamin RJ, Wesley PK, Buxton SE, Garrett TP, Clayberger C, Krensky AM, Norment AM, Littman DR, Parham P (1990) A binding site for the T-cell co-receptor CD8 on the alpha 3 domain of HLA-A2. Nature 345: 41–46

Sands JF, Nikolic-Zugic J (1992) T cell-specific protein-DNA interactions occurring at the CD4 locus: identification of possible transcriptional control elements of the murine CD4 gene. Int Immunol 4: 1183–1194

Sawada S, Littman DR (1991) Identification and characterization of a T cell-specific enhancer adjacent to the murine CD4 gene. Mol Cell Biol 11: 5506–5515

Sawada S, Littman DR (1993) A heterodimer of HEB and an E12-related protein interacts with the CD4 enhancer and regulates its activity in T-cell lines. Mol Cell Biol 13: 5620–5628

Sawada S, Scarborough JD, Killeen N, Littman DR (1994) A lineage-specific transcriptional silencer regulates CD4 gene expression during T lymphocyte development. Cell 77: 917–929

Schonrich G, strauss G, Muller KP, Dustin L, Loh DY, Auphan N, Schmitt-Verhulst AM, Arnold B, Hammerling GJ (1993) Distinct requirements of positive and negative selection for selecting cell type and CD8 interaction. J Immunol 151: 4098–4105

Seong RH, Chamberlain JW, Parnes JR (1992) Signal for T-cell differentiation to a CD4 cell lineage is delivered by a CD4 transmembrane region and/or cytoplasmic tail. Nature 356: 718–720

Siu G, Wurster AL, Lipsick JS, Hedrick SM (1992) Expression of the CD4 gene requires a Myb transcription factor. Mol Cell Biol 12: 1592–1604

Siu G, Wurster AL, Duncan DD, Soliman TM, Hedrick SM (1994) A transcriptional silencer controls the developmental expression of the CD4 gene. EMBO J 13: 3570-3579

Spits H (1994) Early stages in human and mouse T - cell development. Curr Opin Immunol 6: 212–221

Swain SL (1983) T cell subsets and the recognition of the MHC class. Immunol Rev 74: 129 –142

Swat W, Dessing M, von Boehmer H, Kisielow P (1993) CD69 expression during selection and maturation of CD4+CD8+ thymocytes. Eur J Immunol 23: 739–746

Takahama Y, Singer A (1992) Post-transcriptional regulation of early T cell development by T cell receptor signals. Science 258: 1456–1462

Tite JP, Sloan A, Janeway CAJ (1986) The role of L3T4 in T cell activation: L3T4 may be both an Ia-binding protein and a receptor that transduces a negative signal. Mol Cell Immunol 2: 179–190

Travis A, Amsterdam A, Belanger C, Grosschedl R (1991) LEF-1, a gene encoding a lymphoid-specific protein with an HMG domain, regulates T-cell receptor alpha enhancer function. Genes Dev 5: 880–894

Turka LA, Schatz DG, Oettinger MA, Chun JJ, Gorka C, Lee K McCormack WT, Thompson CB (1991) Thymocyte expression of RAG-1 and RAG-2: termination by T cell receptor cross-linking. Science 253: 778–781

van Meerwijk JPM, Germain RN (1993) Development of mature CD8+ thymocytes: selection rather than instruction? Science 261: 911–915

Veis DJ, Sentman CL, Bach EA, Korsmeyer SJ (1993) Expression of the Bcl-2 protein in murine and human thymocytes and in peripheral T lymphocytes. J Immunol 151: 2546–2554

von Boehmer H (1994) Positive selection of lymphocytes. Cell 76: 219–228

Wallace VA, Rahemtulla A, Timms E, Penninger J, Mak TW (1992) CD4 expression is differentially required for deletion of MLS-1a-reactive T cells. J Exp Med 176: 1459–1463

Wassmer P, Chan C, Logdberg L, Shevach EM (1985) Role of the L3T4-antigen in T cell activation. II. Inhibition of T cell activation by monoclonal anti-L3T4 antibodies in the absence of accessory cells. J Immunol 135: 2237–2242

Weintraub H, Davis R, Tapscott S, Thayer M, Krause M, Benezra R, Blackwell TK, Turner D, Rupp R, Hollenberg S (1991) The myoD gene family: nodal point during specification of the muscle cell lineage. Science 251: 761–766

Wu L, Scollay R, Egerton M, Pearse M, Spangrude GJ, Shortman K (1991) CD4 expressed on earliest T-lineage precursor cells in the adult murine thymus. Nature 349: 71–74

Wurster AL, Siu G, Leiden JM, Hedrick SM (1994) Elf-1 binds to a critical element in a second CD4 enhancer. Mol Cell Biol 14: 6452–6463

Xu H, Littman DR (1993) A kinase-independent function of Lck in potentiating antigen-specific T cell activation. Cell 74: 633–643

Zúñiga-Pflücker JC, Jones LA, Longo DL, Kruisbeek AM (1990) CD8 is required during positive selection of CD4⁻/CD8⁺ T cells. J Exp Med 171: 427–437

# Endocytic and Exocytic Regulation of CD4 Expression and Function

M. MARSH and A. PELCHEN-MATTHEWS



Medical Research Council Laboratory for Molecular Cell Biology, and Department of Biochemistry, University College London, Gower Street, London, WCIE 6BT, UK

# 1 Introduction

CD4 is one of the plethora of glycoprotein and glycolipid antigens expressed on the surfaces of lymphocytes (BARCLAY et al. 1993). Many of these antigens are involved in the development and control of lymphocyte function and, as a consequence, their expression at the cell surface must be tightly regulated. Regulation can occur either at the levels of transcription and translation (see, for example, SAWADA et al. 1994) or, alternatively, through the posttranslational activities of the exocytic and endocytic pathways. Here we discuss the processes that operate posttranslationally. In particular we review the interaction of CD4 with the endocytic pathway and discuss how the endocytic properties of CD4 may: (1) function in regulating T cells, and (2) relate to broader issues of membrane trafficking.

The high level of structural, functional and cell biological knowledge of CD4 results from its identification as a key component in the infectious cycle of primate immunodeficiency viruses (HIV-1, HIV-2 and SIV; see SATTENTAU and WEISS 1988; ROBEY and AXEL 1990). Not only do these viruses use CD4 to facilitate their entry into cells, they also employ multiple strategies to reduce the expression of the receptor at the cell surface. This reduction is achieved through changes in the exocytic and endocytic trafficking of CD4 and appears to have a significant influence on the pathology of the diseases induced by these viruses.

# 2 CD4 Expression at the Cell Surface

## 2.1 Expression

CD4 is expressed primarily on differentiating thymocytes and the subset of T lymphocytes which are major histocompatability class II (MHC-II)-restricted and in the main have helper/inducer activities (JANEWAY 1989; ROBEY and AXEL 1990; and elsewhere in this volume). The primary function of CD4 is that of a coreceptor with the T cell antigen receptor/CD3 complex (TcR). By binding to nonpolymorphic regions on the $\beta_2$-domains of MHC-II antigens expressed on thymic epithelial cells or antigen presenting cells (CAMMAROTA et al. 1992; KÖNIG et al. 1992), CD4 is able to augment both the interaction of specific TcRs with the same MHC-II molecule and the signalling activities of that TcR (see elsewhere in this volume). CD4 is also expressed at low levels on T cell precursors in bone marrow and the thymus and, in humans and rats but not in mice, on cells of the monocyte/macrophage lineage, including macrophages, bone marrow-derived dendritic cells, granulocytes and eosinophils. The function of CD4 on these cells is unclear. However, in primates the CD4 molecules expressed on monocytes and lymphocytes can be used by HIV-1, HIV-2 and SIV to productively infect cells (CLAPHAM et al. 1987; SATTENTAU and WEISS 1988; KNIGHT et al. 1990).

## 2.2 Structure and Topology

The CD4 molecule is a nonpolymorphic type I integral membrane glycoprotein. The structure of the protein is illustrated in Fig. 1 and discussed in detail elsewhere in this volume. The primary structure of human CD4, derived from the cDNA sequence, indicates a protein of 458 amino acids of which the first 25 residues form a cleavable signal sequence (MADDON et al. 1985; LITTMAN et al. 1988). The ectodomain, comprising the first 371 amino acids of the mature molecule, is folded into four tandem immunoglobulin-like domains (D1-D4) that are believed to form a 130Å rod-like projection from the membrane (KWONG et al. 1990; RYU et al. 1990; WANG et al. 1990; BRADY et al. 1993). The structure of the ectodomain is described in detail elsewhere in this volume. Here it suffices to mention that the determinants for interaction with MHC-II are located primarily in domains D1 and D2, while residues encompassing the CDR-2 like loop of D1 form the binding site for the HIV-1 envelope glycoprotein, gp120 (See ARTHOS et al. 1989; RYU et al. 1990; WANG et al. 1990; FLEURY et al. 1991). The ectodomain is linked to the membrane by a single 24 amino acid hydrophobic transmembrane domain and is retained in the bilayer by a putative stop transfer sequence of several basic amino acids that immediately follow the transmembrane domain. The 38 COOH-terminal amino acids form the cytoplasmic domain of the molecule and are highly conserved in the species sequenced to date (Fig. 1B).

A

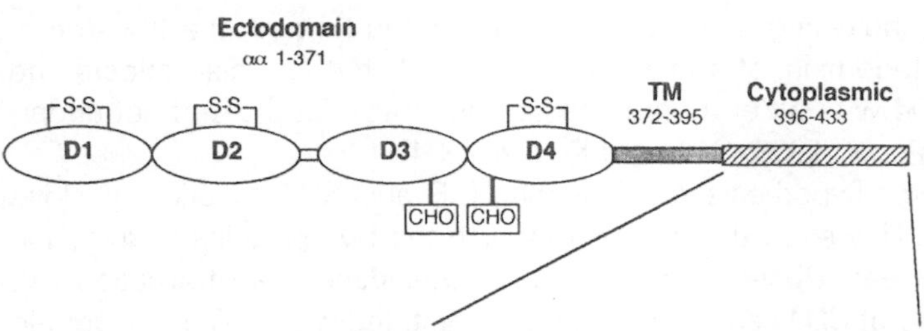


B

```
               393   400       410       420       430
                |     |         |         |         |      % identity with
CD4                                                           human CD4
human          FCV RCRHRRRQAERMSQIKRLLSEKKTCQCPHRFQKTCSPI
macaque        --- --------------------------------------        100
chimpanze      --- ---------Q----------------------------         97
mouse          C-- ----QQ---A--------------------M--SHNL-         79
rat            --- ----QQ---A-----------------S--M--SHNL-         76
rabbit         C-- K-----H--Q------K-------------L---YNLL         76
dog            C-- K-WR-----------------------S--I-----L-         84
cat            C-- K-W------A---H-------------S--L---N--          82
```

**Fig. 1A,B.** Structure of human CD4. **A.** The structure of human CD4. D1–D4 identify the four immunoglobulin domains in the ecto domain of the molecule; *S-S*, indicates the location of disulphide bridges, and *CHO* the sites of the N-linked carbohydrate chains. **B.** The sequences of the cytoplasmic domains of CD4 from various species sequenced to date. The degree of amino acid identity to human CD4, within the 38 amino acid cytoplasmic tail, is indicated

## 2.3 Cell Surface Organisation

Studies of the isolated protein largely support the view that the CD4 protein expressed at the cell surface is monomeric. Recombinant soluble CD4 (sCD4) crystallizes in an associated state (KWONG et al. 1990). However, there is little indication from crystallographic or hydrodynamic analyses (affinity cut-off of about $4 \times 10^5\ M^{-1}$) that isolated CD4 molecules form stable oligomers (DAVIS et al. 1990; KWONG et al. 1990).

Nevertheless, there is indirect evidence that membrane-associated CD4 molecules may form homo-(see KWONG et al. 1990) or hetero-oligomers under certain conditions. The lateral mobility of cell surface CD4 has been analysed by fluorescence recovery after photo bleaching (FRAP). Using FITC-labelled recombinant HIV-1 gp120, a diffusion coefficient of $5–7 \times 10^{-10}$ $cm^2/s$ was measured on CEM cells (PAL et al. 1991); significantly, in the presence of the anti-CD4 monoclonal antibody (Mab) OKT4 the mobile fraction was reduced, suggesting that CD4 molecules may have been clustered on the cell surface. In similar studies using a fringe pattern rather than spot photo bleaching on cells labelled with the rhodamine-conjugated anti-CD4 Mab Q4120 the diffusion constants measured were an order of magnitude slower ($5–9 \times 10^{-11}$ $cm^2/s$) on both T cells and CD4-transfected non-lymphoid cells (A. P-M., J. Davoust and M.M. unpublished). This relatively low mobility suggests that the diffusion of CD4 molecules in the plasma membrane is restricted. Similar diffusion coefficients were measured with full length CD4, in the presence or absence of $p56^{lck}$ (see below), and with CD4 mutants from which the bulk of the cytoplasmic domain has been deleted ($CD4^{cyt-}$), indicating that the mobility is limited through the transmembrane and/or ectodomain of the CD4 molecule. Whether this reflects the association of CD4 with other membrane components is at present not understood (see ZHANG et al. 1993; SAKO and KUSUMI 1994).

The coreceptor hypothesis predicts that TcR and CD4 function in close spacial proximity. However, the degree to which the two proteins form stable complexes is unclear. Observations that anti-CD3-induced cap formation can redistribute CD4, that CD4 can be coimmunoprecipitated with TcR under certain conditions (BURGESS et al. 1991; BEYERS et al. 1992), and that TcR and CD4 can be coordinately down-modulated, suggest that physical interaction can occur. Nevertheless, the association may only occur following T cell activation. Certain anti-TcR antibodies appear to promote the interaction, possibly by inducing conformational changes in TcR that unmask CD4 binding sites (SAIZAWA et al. 1987; JANEWAY 1989), alternatively the association may be facilitated through ligation of MHC-II antigens (BROWN et al. 1993). Analysis of TcR and CD4 by fluorescence resonance energy transfer (FRET) indicate that antibody probes for the two molecules come into close proximity during antibody-induced T cell activation (MITTLER and HOFFMAN 1989). The comodulation of TcR and CD4 may be explained by similarities in the endocytosis signals of the two proteins (see below), thus obviating the requirement for a physical interaction for co-internalization. Furthermore, the observation that TcR and CD4 follow different intracellular itineraries (RUEGG et al. 1992) indicates either that complexes must dissociate following internalization or that CD4 and TcR are internalized independently.

### 2.4 Interaction with p56^lck

In T cells CD4 is associated with p56^lck, a cytoplasmic, membrane-associated, 56 kDa member of the Src family of nonreceptor protein tyrosine kinases expressed primarily in T cells (MARTH et al. 1985; VORONOVA and SEFTON 1986; RUDD et al. 1988; but see TAIEB et al. 1993). The interaction with p56^lck confers a signalling capacity to CD4. Ligation of CD4 with anti-CD4 antibodies activates the kinase activity of p56^lck (VEILLETTE et al. 1989; see elsewhere in this volume), and the activation of p56^lck kinase activity can have either positive or negative effects on T cell activation (TITE et al. 1986; NEWELL et al. 1990; GLAICHENHAUS et al. 1991; A. WEISS 1993). p56^lck may have additional functions: The findings that interaction with p56^lck is required for the close association of CD4 with TcR observed during antibody-induced T cell activation (COLLINS et al. 1992) and that kinase inactive forms of p56^lck can function in p56^lck-dependent responses to antigen (XU and LITTMAN 1993) suggest a structural role for the protein.

## 3 CD4 Synthesis and Transport to the Cell Surface

### 3.1 Synthesis and Transport

From the extensive analysis of the synthesis of other membrane glycoproteins, CD4 is presumed to undergo synthesis on membrane-bound polysomes and cotranslational insertion into the membrane of the rough endoplasmic reticulum (ER). During this process the signal peptide is removed by signal peptidase and high mannose N-linked oligosaccharides are added at consensus Asn-X-Thr sites on D3 and D4 (CD4 does not contain O-linked oligosaccharides). Transport of newly synthesised CD4 has been studied most extensively in HeLa cells using the vaccinia virus bacteriophage T7 polymerase system to express CD4 from transfected cDNA. In this system, transport from the ER to the cis/medial Golgi occurs with a $t_{1/2}$ of 45–60 min, as judged by the acquisition of endoglycosidase H (Endo H) resistance (CRISE et al. 1990; CRISE and ROSE 1992b). The delay in transport to the Golgi is presumably due to the time required for correct folding of the Ig-like domains. CD4 assembly in the ER has not been analysed extensively but is presumably monitored by quality control mechanisms similar to those described for other membrane proteins (HELENIUS 1994). On the majority of CD4 molecules Endo H resistance is conferred by the modification of one N-linked oligosaccharide to complex forms (CRISE et al. 1990; DWEK et al. 1993), though both oligosaccharides may be modified on a fraction of molecules (JABBAR and

NAYAK 1990). Following modification in the Golgi, CD4 appears to move rapidly to the cell surface (RHEE and MARSH 1994).

CD4 is also posttranslationally acylated. Palmitic acid is linked through thioester bonds to cysteines at positions 394 and 397 (CRISE and ROSE 1992b). These reactions are believed to occur either in the ER, intermediate compartment or early Golgi, though palmitoylation at other cellular locations cannot be excluded. The role of cysteine acylation on CD4 is unclear. Mutation of Cys-394 or-397, or both, to Ser does not affect either the rate of transport of newly synthesised CD4 from the ER to the Golgi nor does it affect the ability of CD4 to bind $p56^{lck}$. However, these modified molecules have not been reconstituted into systems in which CD4-dependent activation functions can be examined. Acylation occurs only on cysteines at the junction of the transmembrane and cytoplasmic domains and not on three other cysteine residues in the cytoplasmic domain.

## 3.2 Association with $p56^{lck}$

CD4 associates with $p56^{lck}$ through interaction of its cytoplasmic domain with the unique $NH_2$-terminal domain of the kinase (SHAW et al. 1989, 1990; TURNER et al. 1990). As with other members of the Src family, $p56^{lck}$ is myristoylated at an $NH_2$-terminal glycine (MARCHILDON et al. 1984). In additon, $p56^{lck}$ has palmitate added, through thioester linkages, primarily to the cysteine at position 3 (position 2 of the mature protein) and, to a lesser extent, to Cys-5 (position 4 in the mature protein) (PAIGE et al. 1993; SHENOY-SCARIA et al. 1994). Myristoylation probably occurs cotranslationally or soon after synthesis of the nascent protein is complete (RESH 1990) and appears to be required for efficient targeting of $p56^{lck}$ to membranes where both palmitoylation and interaction with CD4 occur. $p56^{lck}$ in which Gly-2 is mutated does not undergo myristoylation or palmitoylation (SHENOY-SCARIA et al. 1994) and the isolated protein exhibits little if any binding to CD4 (SHAW et al. 1990; TURNER et al. 1990), though limited association may occur in intact cells (A. Pelchen-Matthews and M.Marsh, unpublished). Mutant forms of $p56^{lck}$ in which Cys-2 and Cys-4 are mutated to Ser show reduced binding to CD4, suggesting that palmitate may stabilise the interaction with membranes and/or CD4 (TURNER et al. 1990; SHENOY-SCARIA et al. 1994). Whether palmitoylation is required for the CD4-associated functional activities of $p56^{lck}$ remains to be established but, whereas myristate is stably coupled, $p56^{lck}$-linked palmitate is turned over (PAIGE et al. 1993).

Two additional cysteines in $p56^{lck}$ (Cys20 and Cys23) are essential for interaction with CD4 (SHAW et al. 1990; TURNER et al. 1990). These cysteines are not acylated and mutation of either to Ser or Ala completely disrupts association with CD4. Similarly, two cysteines in the cytoplasmic domain of CD4 (Cys-420 and Cys-422) are required for binding $p56^{lck}$ (SHAW et al. 1990; TURNER et al. 1990). Mutation of either one, or both, to Ser or Ala completely disrupts the physical and functional interaction between CD4 and $p56^{lck}$. In addition to CD4, $p56^{lck}$ is known to interact with high stoichiometry with at least two other membrane proteins, CD8, the TcR coreceptor on MHC class I-restricted T cells, and 4-1BB, a nerve growth factor (NGF) receptor superfamily member expressed on activated T cells (BARBER et al. 1989; KIM et al. 1993). As with CD4, interaction with $p56^{lck}$ requires C-X-C sequences in the cytoplasmic domains of both the α-chain of the CD8 heterodimer and 4–1BB, and Cys20/Cys23 in $p56^{lck}$ (SHAW et al. 1990; TURNER et al. 1990; KIM et al. 1993). These interactions may be stabilised through the co-ordination of a metal ion by the four Cys residues (SHAW et al. 1990). Significantly, the C-X-C sequence in CD8 is located closer to the transmembrane domain than the corresponding motifs in CD4 and 4–1BB and may restrict $p56^{lck}$ binding to CD8.

Additional elements in the cytoplasmic domain of CD4 also play a role in the interaction with $p56^{lck}$. There are several basic amino acids $NH_2$-terminal to the C-X-C sequence in CD4, while the sequence $NH_2$-terminal to the C-X-X-C sequence in $p56^{lck}$ is basic, suggesting that ionic interactions may augment binding to CD4 (TURNER et al. 1990). In addition, phosphorylation of cytoplasmic domain S408 can displace bound $p56^{lck}$ (see below). How and where the CD4/$p56^{lck}$ complex is assembled in the cell and whether assembly is required for efficient transport of $p56^{lck}$ to the plasma membrane remains to be established. When CD4 is confined to the ER, by co-expression with HIV gp160, $p56^{lck}$ is also located on the ER membrane suggesting that CD4 and $p56^{lck}$ may initially associate at the level of the ER and move to the cell surface together (Crise and Rose 1992a).

$p56^{lck}$ also appears to associate at low stoichiometry with a number of other cell surface receptors including the interleukin-2 (Il-2) receptor (HATAKEYAMA et al. 1991), CD5 (RAAB et al. 1994) and various glycophosphatidylinositol (GPI)-linked membrane proteins (see BROWN 1993). The binding in these cases is different to that with CD4 and does not involve cysteine-containing motifs.

# 4 CD4 Endocytosis

## 4.1 CD4 Endocytosis in Immune Function

Although the normal functions of CD4 require its expression at the plasma membrane, the stimuli that lead to T cell activation also induce CD4 down-regulation. The cell surface expression of other T cell antigens is also modulated under these conditions. For example the CD3/TcR complex, which is regulated by similar endocytosis signals on the cytoplasmic domains of the CD3 γ and δ chains (see below), is also down-modulated in response to stimulation, while the expression of other molecules such as CD2 can be up-regulated. One function of the modulation of surface components may be to facilitate release of T cells from antigen presenting cells following activation (ROSENSTEIN et al. 1990). In addition, cells which have been activated through encounter with an antigen presenting cell may become transiently desensitized or refractory to further stimulation. CD4 down-modulation may also underlie the profound and long-term energy and immune suppression observed after treatment with anti-CD4 antibodies in vivo (see elsewhere in this volume). Although CD4 function at the cell surface may be down-modulated by ligand-mediated occlusion of crucial sites on the molecule, the principal mechanism for regulating expression is the removal of cell surface molecules through endocytosis.

## 4.2 The Endocytic Pathway

The endocytic pathway is a highly dynamic membrane trafficking system that operates in virtually all nucleated cells. Endocytosis occurs primarily through the continual invagination of plasma membrane clathrin-coated pits to form coated vesicles. This pathway accounts for the majority of the constitutive endocytic activity in several cell types (WATTS and MARSH 1992) and is responsible for : (1) fluid phase endocytosis, (2) the rapid receptor-mediated endocytosis of specific cell surface proteins and their ligands, and (3) the bulk flow uptake and recycling of plasma membrane. Cells may internalize and recycle an area equivalent to the entire cell surface every 30–120min. Clathrin-coated pits occupy about 2% of the plasma membrane area (GRIFFITHS et al. 1989; PELCHEN-MATTHEWS et al. 1991; but see MILLER et al. 1991) and form coated vesicles at a rate of ~1000 per cell per min in, for example HeLa or baby hamster kidney (BHK) cells (MARSH and HELENIUS 1980; PELCHEN-MATTHEWS et al. 1991). In lymphoid cell lines such as CEM and SupT1, which have a cell surface area and cytoplasmic volume ~15%–20% that of a BHK cell, the formation of ~100 vesicles/cell min accounts for an equivalent endocytic capacity (PELCHEN-MATTHEWS et al. 1991).

The structure of coated pits and characterization of their components has recently been reviewed (ROBINSON 1992; SCHMID 1992). Briefly, proteins of the AP-2 adaptor complex bind to the inner surface of the plasma membrane and recruit clathrin triskelions, thus forming the characteristic polygonal lattice of coated pits. The adaptor proteins also have the capacity to interact with structural elements or signals in the cytoplasmic domains of certain integral membrane proteins and thus concentrate these proteins in invaginating coated pits. For constitutively internalized receptors, such as the low density lipoprotein and transferrin receptor (LDL-R, Tf-R), the endocytosis signals frequently contain a tyrosine residue within short sequences of about six amino acids that are believed to form tight β-turns (VAUX 1992; TROWBRIDGE et al. 1993). Other signals for association with coated pits must also exist (see below), although these are at present less well defined.

Coated pits pinch off from the plasma membrane, are uncoated and subsequently fuse with early endosomes. Within the early endosomal system, internalized membrane components, receptors and their cognate ligands are sorted for distribution to different sites in the cell. A large proportion of the membrane and many receptors are recycled to the cell surface. Certain specific receptors, e.g. ligated epidermal growth factor receptors (EGF-Rs) or antigen cross-linked Fc receptors, are targeted to lysosomes for degradation. Other routes from early endosomes to transcytotic vesicles, the Golgi complex, synaptic vesicles, or other endosomal compartments, e.g. late endosomes and compartments involved in loading peptides onto MHC-II antigens, have been identified in various cell types. The mechanisms involved in sorting membrane components to these different compartments is as yet unclear. Acidification of endosomes by vacuolar $H^+$-ATPases is involved to some extent (MELLMAN et al. 1986) while coat proteins functionally analogous to the clathrin adaptor proteins may mediate other aspects of sorting.

## 4.3 Measurement of CD4 Endocytosis

For CD4, the terms "endocytosis" and "down-regulation" are frequently considered equivalent. In practice, down-regulation of CD4 involves a series of events which together reduce the surface expression of the protein. Endocytosis is an essential part of this process but, despite the fact that changes in the rate of CD4 endocytosis can potentially account for down-modulation (see below; PELCHEN-MATTHEWS et al. 1993), endocytosis alone is not sufficient. Assays which monitor the cell surface expression of CD4, such as antibody binding and FACS analysis, provide useful information on the occurrence and extent of down-regulation, but cannot give accurate quantitative and kinetic data for endocytosis and recycling. Measurement of endocytosis requires methods which follow molecules from the plasma membrane into the cells. Thus, cell surface CD4 must be identified by labelling with, e.g. radioiodinated antibody tracers, or by covalent modification using conditions (e.g. 4°C) that inhibit endocytosis (PELCHEN-MATTHEWS et al. 1989, 1991). When the labelled cells are then warmed to temperatures permissive for endocytosis (e.g. 37°C), tagged CD4 molecules may be internalized. The amount of CD4 uptake can then be measured by returning the cells to 4°C, removing any label remaining at the cell surface (by acid washing or proteolytic digestion for example), and quantitating the intracellular CD4 pool by $\gamma$-counting or immunoprecipitation. The assays can be further refined to monitor the degradation and recycling of internalized antigen. In addition, the distribution and uptake of CD4 can be visualized directly by immunofluorescence or immunoelectron microscopy using appropriate reagents.

## 4.4 Constitutive CD4 Endocytosis in T Cells

The constitutive endocytosis of CD4 has been studied on a variety of leukaemia or lymphoma-derived T cell lines and in peripheral blood lymphocytes using both radioiodinated antibody and cell surface biotinylation internalization assays. In all cases, CD4 uptake was negligible (Table 1), usually with endocytosis rates of <0.5%/min and <10% of the cell surface antigen accumulating in the cells after 1–2 h at 37°C (PELCHEN-MATTHEWS et al. 1991). In keeping with this observation, immunoelectron microscopy revealed that cell surface CD4 was distributed over regions of noncoated plasma membrane and microvilli and was excluded from

**Table 1.** Kinetics of the constitutive endocytosis of CD4 and mutated CD4 molecules, and comparison with transferrin receptor endocytosis

| Cell line | Number of experiments | Rate of endocytosis (%/min) | % internal equilibrium |
|---|---|---|---|
| CD4 endocytosis in lymphoid cells | | | |
| Purified peripheral blood lymphocytes | 1 | 0.64 | 8.2 |
| SupT1 | 2 | 0.3 ± 0.2 | 6.8 ± 2.8 |
| CEM | 2 | 0.2 ± 0.03 | 9.6 ± 6.9 |
| Jurkat | 2 | 0.4 ± 0.05 | 4.4 ± 0.2 |
| MOLT–4 | 1 | 0.2 | 4.2 |
| CD4-transfected RPMI cells | 3 | 0.5 ± 0.45 | 13.4 ± 0.6 |
| CEM cells transfected with: | | | |
| $CD4^{-cyt399}$ | 1 | 1.30 | 29 |
| $CD4^{-cyt401}$ | 1 | 0.91 | 22 |
| CD4 endocytosis in nonlymphoid cells | | | |
| HeLa-CD4 | 8 | 1.8 ± 0.9 | 36 ± 8 |
| NIH3T3-CD4 | 20 | 3.91 ± 1.21 | 41 ± 10 |
| HL-60 [a] | 3 | 4.8 ± 1.2 | 49 ± 8 |
| U937 [a] | 2 | 2.7 ± 0.6 | 29 ± 4 |
| HeLa cells transfected with: | | | |
| $CD4^{4cyt-}$ | 6 | 0.69 ± 0.22 | 18 ± 3 |
| CD4(S408A) | 5 | 0.79 ± 0.09 | 32 ± 3 |
| CD4(C420S) | 3 | 3.12 ± 1.42 | 42 ± 7 |
| CD4(C420, 422S) | 1 | 0.84 | 34 |
| Kinetics of transferrin receptor internalization | | | |
| CEM | 1 | 7.2 | 80 |
| HeLa-CD4 | 2 | 11.4 ± 0.9 | 86 ± 5 |
| NIH-3T3-CD4 | 2 | 9.5 ± 2.8 | 75 ± 7 |

CD4 endocytosis was measured using radioiodinated anti-CD4 monoclonal antibodies (Leu3a or Q4120) or radioiodinated transferrin as described in PELCHEN-MATTHEWS et al. (1989, 1991, 1992, see also text)

[a]Comparable results were obtained in the presence of an excess of isotype-matched control monoclonal antibody to block possible binding of the anti-CD4 to Fc receptors

coated pits and coated vesicles (PELCHEN-MATTHEWS et al. 1991). Despite their low levels of CD4 endocytosis, these cells have an endocytic capacity equivalent to that of HeLa or BHK cells (see above) and can internalize transferrin at rates similar to those observed in nonlymphoid cells (Table 1).

## 4.5 Constitutive CD4 Endocytosis in Nonlymphoid Cells

In contrast to the results in lymphoid cells, nonlymphoid cells expressing CD4 constitutively endocytose the molecule (Table 1). CD4 internalization has been demonstrated in the promyelocytic cell line HL-60 (PELCHEN-MATTHEWS et al. 1991), in U937 cells, and in HeLa, NIH3T3 and CHO cells expressing CD4 from transfected cDNAs (SUZUKI et al. 1991; KELLER et al. 1992; PELCHEN-MATTHEWS et al. 1989, 1992). The rates of uptake typically range from 2 to 4% of the cell surface pool per min (Table 1). CD4 endocytosis is balanced by recycling back to the cell surface (PELCHEN-MATTHEWS et al. 1989), so that in antibody endocytosis assays a plateau is reached in the accumulation of endocytosed tracer (see Fig. 2). At

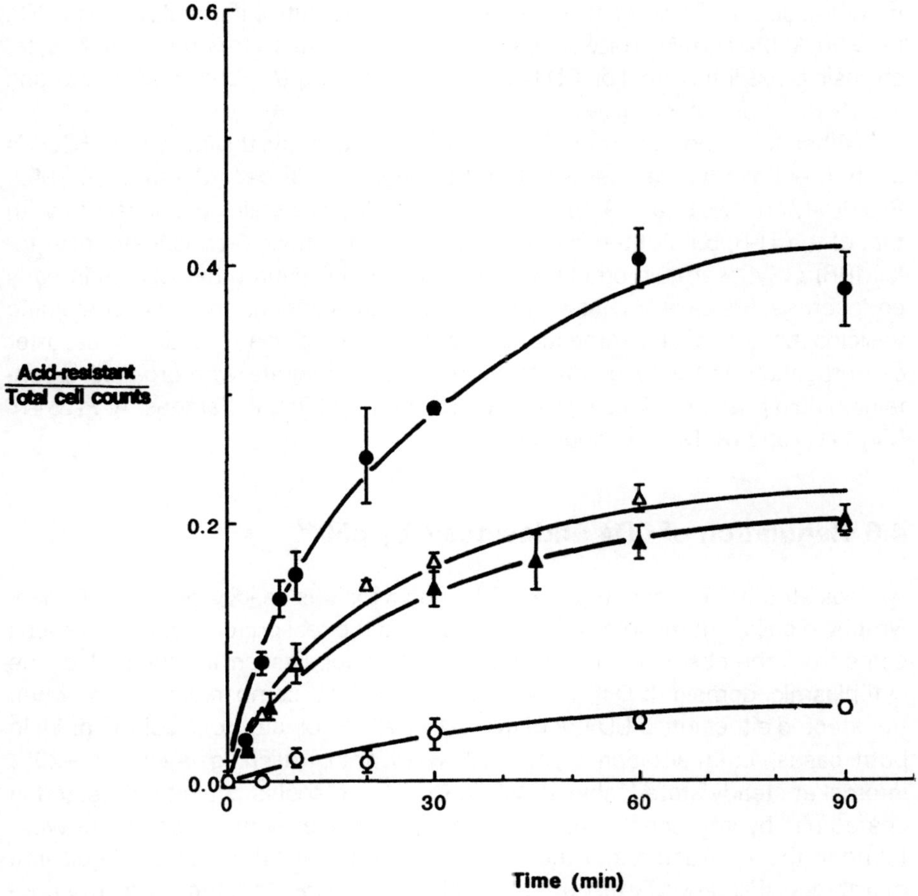


**Fig. 2.** Constitutive endocytosis of CD4 in lymphoid and nonlymphoid cells. CD4 was labelled at 4°C with 0.3 *nM* $^{125}$I-labelled anti-CD4 monoclonal antibody before cells were washed and incubated at 37°C for various times. Cells were then cooled and either acid-washed to remove antibody remaining at the cell surface or harvested directly to determine total cell-associated label. The curves show the proportion of acid-resistant (internal) to total cell-associated activity at various times. *Closed symbols*, transfected HeLa cells; *open symbols*, CEM cells. *Circles* show wild-type CD4; *triangles* $CD4^{cyt-}$

steady state (30–60 min after the onset of uptake), 40%–50% of the initial cell surface CD4 is in an intracellular compartment. Comparable levels of CD4 endocytosis are observed in assays using cell surface biotinylation (i.e. in the absence of a ligand tracer) or when CD4 is labelled with Fab' fragments, recombinant HIV gp120, bivalent antibody, or antibody and protein A (PELCHEN-MATTHEWS et al. 1989, 1991; SUZUKI et al. 1991; KELLER et al. 1992), indicating that CD4 uptake is independent of ligand binding. CD4 is internalized through coated pits and coated vesicles. Immunoelectron microscopy shows that CD4 is enriched in these structures by about three fold (KELLER et al. 1992; MARSH et al. 1992; PELCHEN-MATTHEWS et al. 1991, 1992) Furthermore, treatments which disrupt endocytosis through coated pits (e.g. exposure of cells to hypertonic medium, see van Deurs et al. 1989) inhibit the constitutive internalization of CD4 by >90% (Pelchen-Matthews et al. 1993). In contrast, treatment with cytochalasin D has little effect on CD4 uptake and recycling (A. Pelchen-Matthews and M. Marsh unpublished).

When followed by immunofluorescence microscopy, the internalized CD4 is seen in endosomal vesicles scattered throughout the cell (Marsh et al. 1992; Pelchen-Matthews et al. 1993). In HeLa-CD4 cells, this staining colocalises with that of the Tf-R, but not the cation-independent mannose 6-phosphate receptor (ciMPR) or lysosomal membrane glycoproteins, indicating that CD4 is in early endosomes (Pelchen-Matthews et al. 1993). In addition, the CD4-containing vesicles have the characteristic bouyant density of early endosomes after centrifugation on Percol gradients, where they comigrate with organelles containing fluid phase markers internalized for 5 min at 37°C (I. Parsons, A. Pelchen-Matthews and M. Marsh unpublished.).

## 4.6 Regulation of CD4 Endocytosis by $p56^{lck}$

As indicated above, constitutive CD4 endocytosis is readily detected in nonlymphoid cells, but minimal in T cells. A clue to the reason for these differences came from the observation that mutant CD4 molecules lacking the bulk of the cytoplasmic domain ($CD4^{cyt-}$) were internalized to comparable levels when transfected either into a CD4-negative CEM cell line or into HeLa cells (Fig. 2). In both cases, internalization was ~ 0.8%–1%/min, reaching levels of ~20% internal at steady state (Table 1). While $CD4^{cyt-}$ molecules could be detected in coated pits by immunoelectron microscopy, they were not enriched, but were found in these structures at the same density as over the rest of the plasma membrane (Pelchen-Matthews et al. 1991). This data, together with morphometric analysis of the endocytic compartments, demonstrates that the $CD4^{cyt-}$ constructs are internalized by bulk flow endocytosis (Pelchen-Matthews et al. 1991). The augmented levels of endocytosis of full length CD4 observed in nonlymphoid cells indicate that the CD4 cytoplasmic domain contains an endocytosis signal (see below). In contrast, in lymphoid cells the CD4 cytoplasmic domain mediates the exclusion of the molecule from coated pits and the lack of endocytosis.

Since the cytoplasmic domain of CD4 mediates distinct functions in the different cellular backgrounds, it seemed likely that some T cell specific component modulates the interaction of CD4 with coated pits. An obvious candidate was $p56^{lck}$, which is expressed in T cells (Marth et al. 1985), associates with the cytoplasmic domain of CD4 and is expressed at levels sufficient to form complexes with a high proportion of cell surface CD4 molecules. Furthermore, the association of CD4 with $p56^{lck}$ correlates with a low level of constitutive CD4 endocytosis (Pelchen-Matthews et al. 1992). When $p56^{lck}$ is transfected into CD4-expressing nonlymphoid cell lines, the constitutive endocytosis of CD4 is inhibited, and CD4 molecules are excluded from coated pits as they are on lymphocytes (Pelchen-Matthews et al. 1992). The inhibition of CD4 endocytosis requires a direct interaction of CD4 with $p56^{lck}$ since it is more pronounced the higher the levels of $p56^{lck}$ expressed, is blocked by mutations in the cytoplasmic domain of CD4 or in the unique $NH_2$-terminal region of $p56^{lck}$ that disrupt CD4-$p56^{lck}$ association, and cannot be mimicked by other Src family kinases (Pelchen-Matthews et al.1992). By contrast with its effect on CD4 endocytosis, $p56^{lck}$ does not influence Tf-R or fluid phase endocytosis.

By inhibiting CD4 endocytosis, $p56^{lck}$ restricts CD4 to the plasma membrane. Expression at the cell surface is clearly important for the functional activity of CD4. Nevertheless, in cells that do not express $p56^{lck}$ and which constitutively internalize CD4, approximately 60% of the cycling pool of CD4 is present at the cell surface at any one time. As discussed elsewhere in this volume, $p56^{lck}$ provides a signalling capacity to CD4, and the CD4-$p56^{lck}$ complex is required for efficient activation of T cells. It is not apparent that these functions might be facilitated by preventing CD4 internalization. The observation that $p56^{lck}$ is required for CD4-TcR juxtapositioning (Collins et al. 1992) and that kinase inactive forms of $p56^{lck}$ can augment TcR activation in at least some systems (Xu and Littman 1993) suggests an additional structural role for $p56^{lck}$. The effect of $p56^{lck}$ on CD4 endocytosis may reflect this structural function.

The mechanism by which $p56^{lck}$ prevents CD4 endocytosis is not clear. One possibility is that $p56^{lck}$ masks or disrupts the endocytosis signal present in the cytoplasmic domain of CD4. But this should only reduce the rate of CD4 internalization to that of the $CD4^{cyt-}$ constructs, and the observed effect of $p56^{lck}$ is much greater than this. Another possibility is that the large size of the $p56^{lck}$ protein might sterically prevent CD4 from interacting with coated pits. However, growth factor receptors such as EGF-R, which also have bulky intracellular kinase domains, can dimerize, enter coated pits and undergo endocytosis (Cochet et al. 1988). As a further possibility, $p56^{lck}$ may interact with other cellular components present in both lymphoid and nonlymphoid cells and thereby anchor CD4 on the noncoated plasma membrane. The CD4-$p56^{lck}$ complex has been reported to associate with a number of other proteins involved in signal transduction, but these interactions are of low stoichiometry (Rudd 1990) and unlikely to affect endocytosis. $p56^{lck}$ molecules expressed in NIH3T3 cells in the absence of CD4 have been reported to associate with the cytoskeleton (Louie et al. 1988), and a similar link to the cytoskeleton may exist in lymphocytes. CD4 can be capped on T cells and the capping is sensitive to temperature, metabolic inhibitors and agents such as cytochalasin B, which disrupt the actin-based cytoskeleton. The association of the CD4-$p56^{lck}$ complex with the cortical cytoskeleton would not only provide a mechanism to prevent CD4 internalization but, during activation, might facilitate the relocation of CD4 on the T cell surface to sites of interaction with antigen presenting cells (Podack and Kupfer 1991).

## 4.7 CD4 Down-Regulation After Stimulation of Protein Kinase C

CD4 can be rapidly down-regulated from the surfaces of T cells by a variety of stimuli such as exposure to an appropriate antigen (ACRES et al. 1986; WEYAND et al. 1987; RIVAS et al. 1988) or treatment with cross-linking antibodies against CD4 (COLE et al. 1989), the CD3/TcR complex (RIVAS et al. 1988) or CD2 (BLUE et al. 1989). Most of these stimuli operate through the activation of protein kinase C (PKC) and, accordingly, the most effective triggers for CD4 down-regulation are PKC-activating phorbol esters such as phorbol myristate acetate (PMA) or phorbol dibutyrate (PDB). These reagents have therefore been used extensively to analyse CD4 modulation biochemically (see, for example, HOXIE et al. 1986b, 1988; HURLEY et al. 1989; SLECKMAN et al. 1989; SHIN et al. 1990, 1991; RUEGG et al. 1992; PELCHEN-MATTHEWS et al. 1993).

On both lymphoid and CD4-transfected nonlymphoid cells, PMA or PDB induce the rapid, synchronous and highly reproducible removal of CD4 from the cell surface: Up to 70%-80% may be down-regulated to an intracellular compartment within 30 min. If the cells are maintained in PMA for extended periods, CD4 is gradually re-expressed on the cell surface and, for example in HeLa-CD4 cells, may recover to the levels observed in untreated cells after 24 h. The re-expression of CD4 presumably reflects the phorbol ester-induced down-modulation of PKC (COLLINS and ROZENGURT 1982).

Analysis of the mechanism by which phorbol esters down-regulate CD4 expression has shown that the process is complex and consists of a number of distinct events.

### 4.7.1 CD4 Phosphorylation

Like the encounter with specific antigen (ACRES et al. 1986; BLUE et al. 1988), treatment of CD4-expressing cells with PMA induces a rapid, transient phosphorylation of CD4 (HOXIE et al. 1988b; HURLEY et al. 1989; SHIN et al. 1990). All three of the serine residues in the CD4 cytoplasmic domain (see Fig. 1B) are located within consensus sites for PKC-mediated phosphorylation (KENNELLY and KREBS 1991). Studies with mutated CD4 molecules have indicated that all three sites can be phosphorylated in cells treated with PMA or by purified PKC in vitro (SHIN et al. 1990). However, the level of phosphate incorporated varies (S408 > S415 > S431) in parallel with the relative importance of the residues for down-regulation (SHIN et al. 1990). Other kinases may also regulate CD4 through the phosphorylation of cytoplasmic serine residues (KAMMER et al. 1988). However, the identity of these kinases and their mode of interaction with CD4 remain to be established.

### 4.7.2 Dissociation of CD4 from $p56^{lck}$

Although the constitutive endocytosis and recycling of CD4 on T cells is restricted by $p56^{lck}$, these cells do efficiently modulate their plasma membrane CD4 levels when treated with phorbol esters (CASSEL et al. 1983; HOXIE et al. 1986b, 1988;

SHIN et al. 1991; PELCHEN-MATTHEWS et al. 1993). An indication of how CD4 endocytosis may be possible in the presence of $p56^{lck}$ came from the observation that $p56^{lck}$ is not coprecipitated with CD4 from lysates of PMA-treated T cells (HURLEY et al. 1989). This is due to dissociation of the CD4-$p56^{lck}$ complex, probably as a direct consequence of CD4 phosphorylation (HURLEY et al. 1989). Mutant CD4 molecules in which Ser- 408 is changed to Ala (CD4(S408A)) do not dissociate from $p56^{lck}$ and are poorly down-regulated (BEYERS et al. 1991; SLECKMAN et al. 1992). However, other features are also required as CD4(L413D) mutant molecules appear to undergo proper phosphorylation, yet $p56^{lck}$ does not dissociate (SLECKMAN et al. 1992). The dissociation of $p56^{lck}$ appears both to precede and be essential for CD4 internalization (SLECKMAN et al. 1992; YOSHIDA et al. 1992). This is also apparent in the model system of NIH3T3 cells stably transfected with both CD4 and $p56^{lck}$. When these cells are treated with PMA, the CD4-$p56^{lck}$ complex dissociates with a $t_{1/2}$ of 1–2 min (Fig. 3C); PMA-induced CD4 endocytosis occurs after a brief lag, and significant levels of internalized CD4 are only observed at times (≥10 min) when the majority of $p56^{lck}$ has dissociated (Fig. 3B, see also PELCHEN-MATTHEWS et al. 1993).

### 4.7.3 Increased association of CD4 with Coated Pits and Endocytosis

Dissociation of $p56^{lck}$ is not sufficient to explain the modulation of CD4 observed in lymphocytes. CD4 molecules no longer bound by $p56^{lck}$ are free to enter coated pits and might be expected to behave as they do in non-lymphoid cells; in this case, a new steady state of internalization and recycling would be reached, with about 40% of the CD4 relocating to endosomes. Much greater levels of down-regulation are in fact observed. Furthermore, PMA also induces CD4 down-regulation in nonlymphoid cells (CLAPHAM et al. 1987; SHIN et al. 1991; PELCHEN-MATTHEWS et al. 1993; see for example Fig. 3A).

In the nonlymphoid Hela-CD4 or NIH3T3-CD4 cells, PMA increases the rate of CD4 endocytosis by three to five-fold and doubles the level of internal CD4 at steady state. The endocytosis rates observed in the presence of PMA (10%–13%/min) are comparable to the constitutive internalization of Tf-R or LDL-R (see Table 1, and TROWBRIDGE et al. 1993; HANSEN et al. 1992). The trafficking of $CD4^{cyt-}$ molecules, and fluid phase endocytosis, are not affected by PMA in these cells (PELCHEN-MATTHEWS et al. 1993), indicating that the effect is not due to a general modulation of the endocytic activity by the phorbol ester, as may occur in some cell types (see WATTS and MARSH 1992). Like constitutive CD4 endocytosis, PMA-stimulated endocytosis is inhibited by >90% by treatments which disrupt coated vesicle formation (PELCHEN-MATTHEWS et al. 1993), suggesting that the increased uptake occurs through coated pits. This can be demonstrated directly by electron microscopy. Quantitation of the distribution of anti-CD4/protein A-gold complexes on PMA-treated cells shows that within 2 min of warming there is a two to three fold increase in the density of gold particles in coated pits (PELCHEN-MATTHEWS et al. 1993). It is likely that the increased endocytosis of CD4 is due to the PKC-mediated phosphorylation of the

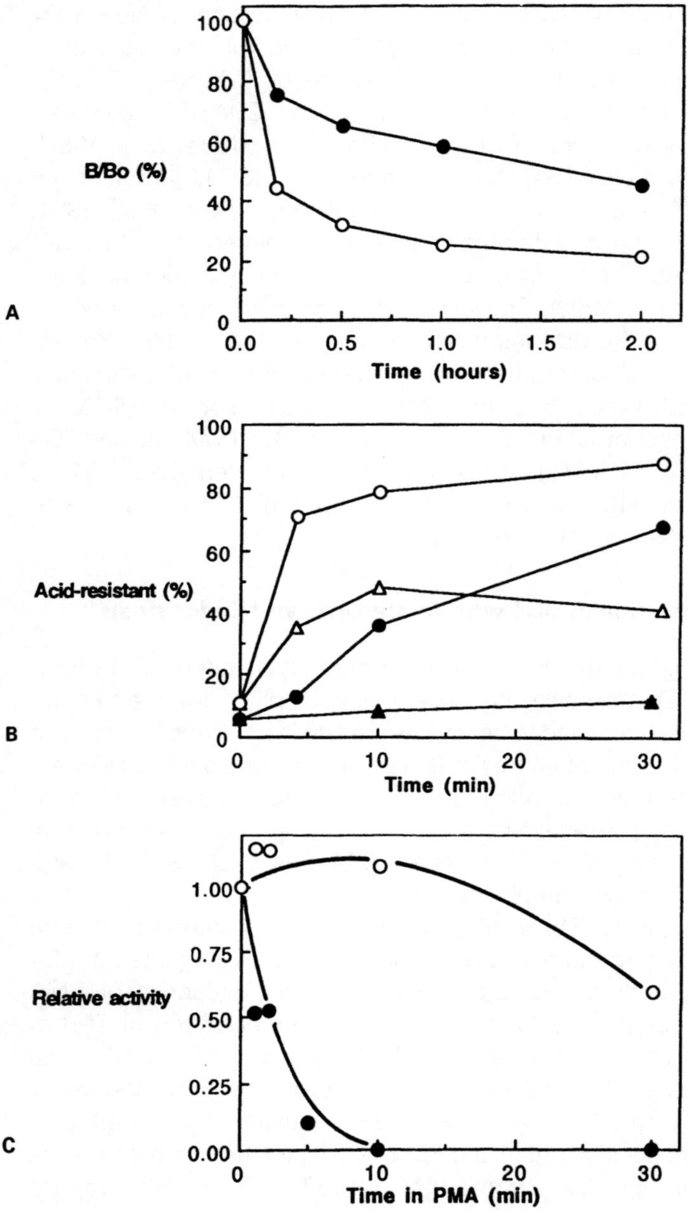


**Fig. 3A-C.** CD4 down-regulation in transfected NIH3T3 cells. **A** Time course of CD4 down-regulation induced by PMA (100 ng/ml) in NIH3T3 cells transfected with CD4 only (*open circles*) or with CD4 and p56$^{lck}$ (*Closed circles*) **B** Endocytosis time courses, determined as in Fig. 2, for NIH3T3 cells transfected with CD4 only (*open symbols*) or with CD4 and p56$^{lck}$ (*closed symbols*). *Circles* show uptake in the presence of PMA; *triangles*, endocytosis under control conditions. **C** Dissociation of CD4 and p56$^{lck}$. NIH3T3 cells transfected with both CD4 and p56$^{lck}$ were treated with 100 ng/ml PMA for various times and then lysed in Tris/NaCl buffer containing 3% NP40. Cell lysates were either immunoprecipitated with anti-CD4 antibodies (*Closed circles*) or analysed directly (*open circles*) by SDS-PAGE and immunoblotting for p56$^{lck}$. Autoradiograms were quantitated by densitometry

cytoplasmic domain of the molecule since the mutant CD4(S408A) molecules respond with a reduced and slower increase in CD4 endocytosis when stimulated with PMA.

Although a phorbol ester-induced increase in the association of CD4 with coated pits has yet to be demonstrated directly in T cells, PMA does increase the rate of CD4 endocytosis observed in lymphoid cell lines by ten to 20-fold (e.g. to 4.2%/min in SupT1 cells) to levels exceeding the rates of constitutive endocytosis in nonlymphoid cells (PELCHEN-MATTHEWS et al. 1993). The effect of PMA on CD4 endocytosis is likely to be greater than this, as asynchronous and/or incomplete dissociation of CD4-p56$^{lck}$ will underestimate the measured rates of CD4 endocytosis. Thus, as in nonlymphoid cells, PMA must also increase CD4 internalization through the coated pits of lymphoid cells.

### 4.7.4 Sorting of Internalized CD4 to Lysosomes

Simple mathematical models which take into account the rates of endocytosis and recycling and the relative sizes of the cell surface and endocytic pools of CD4, indicate that a three to four fold increase in the rate of internalization, in the absence of an effect on recycling, will redistribute the majority of cell surface CD4 to an intracellular pool with similar kinetics to those measured biochemically, i.e. an increase in the rate of internalization alone could account for CD4 down-regulation (PELCHEN-MATTHEWS et al. 1993). CD4 would continue to cycle between the cell surface and endosomes, but the half time on the surface would be reduced. However, biochemical analysis of PMA-treated lymphoid and nonlymphoid cells indicates that internalized CD4 is degraded by acid hydrolases (SHIN et al. 1990; RUEGG et al. 1992).

In PMA-treated HeLa-CD4 cells, altered sorting of internalized CD4 has been demonstrated by immunofluorescence. In these cells, constitutively internalized CD4 is found in early endosomes (see above). Following treatment with PMA, CD4 is initially observed in these structures, but subsequently moves to a perinuclear location. The distribution of CD4 overlaps with that of ciMPR (PELCHEN-MATTHEWS et al. 1993), but is distinct from lysosomal markers. In cells expressing CD4$^{cyt-}$ the internalized antigen remains in early endosomes and does not localise into the perinuclear region. Together these observations indicate that a third effect of phorbol esters on the normal trafficking of CD4 is to reroute internalized antigen from the endosome/plasma membrane recyling pathway and direct it to late endosomes and lysosomes where it is later degraded. The altered sorting occurs in early endosomes, appears to require the cytoplasmic domain of the molecule, and may involve the same PKC-mediated serine phosphorylation reactions as those involved in p56$^{lck}$ dissociation and increased endocytosis.

## 4.8 Signals in the Cytoplasmic Domain of CD4

Studies of transfected cells expressing $CD4^{cyt-}$ molecules demonstrate that the cytoplasmic domain of CD4 is crucial for most, if not all, of the steps involved in the regulation of CD4 cell surface expression, endocytosis and cellular trafficking. As indicated earlier, this domain contains the conserved C-X-C motif required for interaction with $p56^{lck}$. Other amino acids are involved in modulating the interaction with $p56^{lck}$ (see above) and providing the signals that mediate coated pits and endosomal sorting reactions.

Mutational analysis of the CD4 cytoplasmic domain has to date been focused on identifying residues required for CD4 down-regulation by phorbol esters (Fig. 4). These studies have demonstrated the importance of the cytoplasmic serine residues, probably as substrates for PKC. Mutants lacking S408 are only down-regulated by ~20%. S415 makes a lesser, though still significant, contribution, and the effects of mutations at both S408 and Ser-415 appear to be additive. The least conserved serine residue, Ser-431, near the COOH-terminus has a negligible effect. Mutation of all three serine residues completely inhibits CD4 modulation (BEDINGER et al. 1988; SHIN et al. 1990). CD4 mutants truncated after K417 are fully down-regulated by PMA (even though they cannot bind $p56^{lck}$), placing the down-regulation motif in the $NH_2$-terminal half of the cytoplasmic domain. Within this portion of the molecule, the conserved amino acids M407, I410, L413 and L414 are also essential for PMA-induced CD4 modulation (SHIN et al. 1991).

The signals for coated pit localisation and endosomal sorting in CD4 have not yet been mapped directly. The endocytosis signal is likely to be distinct from the well-characterized tyrosine-containing tight turn motifs identified in constitutively internalized receptors (see above). The cytoplasmic domain of CD4 does not contain a tyrosine residue, and there is only one aromatic amino acid (F426) near the COOH-terminus. This portion of the molecule is dispensable for down-regulation and in modelling studies does not appear to form a β-turn.

The endocytosis motif is thus likely to be located in the membrane proximal region of the cytoplasmic domain (R396-K417) identified above. This region of the CD4 molecule has homology with a number of other proteins that undergo regulated endocytosis (e.g. the CD3γ and δ chains, the glucose transporter GLUT 4, and the EGF-R) and may contain a second class of endocytosis motif. Studies of the TcR complex have indicated that a sequence in the cytoplasmic domain of the CD3γ chain, which contains a pair of leucine residues, can act as an endocytosis and lysosomal targeting signal (LETOURNEUR and KLAUSNER 1992). Subsequent analyses indicate that the di-leucine signal requires a nearby serine residue within a PKC consensus sequence (DIETRICH et al. 1994). The dileucine endocytosis signal appears to be active only when the serine is phosphorylated, i.e. it can be switched from an inactive to a functional state by phosphorylation. As indicated earlier, the two leucines in the cytoplasmic domain of CD4 (L413 and L414) are essential for down-regulation and are also situated between serine

| | 396 400 410 420 430 | TPA-induced Down-regulation | References |
|---|---|---|---|
| CD4wt | RCRHRRRQAERMSQIKRLLSEKKTCQCPHRFQKTCSPI | +++ | |
| $CD4^{cyt-}$ (stop399) | ---H | - | 1 |
| $CD4^{cyt-}$ (stop401) | -----R | - | 1 |
| $CD4^{cyt-}$ (stop402) | ------R | - | 2, 3, 5 |
| S408A | ------------A------------------------- | + | 1, 2, 4 |
| S415A | -------------------A------------------ | ++ | 4, 1* |
| S431A | -----------------------------------A-- | +++ | 4 |
| S408/415A | ------------A------A------------------ | +/- | 4 |
| S408/431A | ------------A----------------------A-- | + | 4 |
| S415/431A | -------------------A---------------A-- | ++ | 4 |
| S408/415/431A: | ------------A------A---------------A-- | - | 4 |
| Stop Arg412 | ---------------K | - | |
| Stop Lys418 | ---------------------K | +++ | 5 |
| Stop Phe426 | -----------------------------R | +++ | 5 |
| M407T: | -----------T-------------------------- | - | 5 |
| I410D: | --------------D----------------------- | - | 5 |
| L413N: | -----------------N-------------------- | - | 5 |
| L413D: | -----------------D-------------------- | - | 5 |
| L414D: | ------------------D------------------- | - | 5 |

**Fig. 4.** CD4 cytoplasmic domain mutants. Truncations or changes to one or more amino acids in the cytoplasmic domain of CD4 have been generated and the ability of these mutants to undergo phorbol ester-induced down-regulation examined. Removal of the entire cytoplasmic domain, replacement of all three cytoplasmic serine residues or the hydrophobic residues between M407 and L414 completely abrogates the ability to down-regulate. For some of these mutants the effects on endocytosis are summarised in Table 1. References: *1.* BEDINGER et al. 1988; *2,* MADDON et al. 1988; *3,* SLECKMAN et al. 1989; *4,* SHIN et al. 1990. *5,* SHIN et al. 1991. *Asterisk,* in reference 1, a CD4 (S415L) mutant which behaves similarly is described.

residues (S408 and S415) which can be phosphorylated by PKC. The importance of phosphorylation on these serine residues is indicated by the observation that the impairment of internalization in the CD4(S408A) mutant can be partially restored by a compensating mutation of A404S (S404 is phosphorylated in this mutant, SHIN et al. 1991).

It thus appears that a phosphoserine/di-leucine containing signal mediates entry of CD4 into coated pits. The precise structure of this signal and the mechanisms involved in its recognition by the cellular sorting machinery are at present unclear. The membrane-proximal region of CD4 may assume an amphipathic α-helical conformation, with the hydrophobic residues M407, I410, L413 and L414 arranged on the same side of the helix (SHIN et al. 1991). Phosphorylation, in particular at S408, might modify this structure or reveal conformational features which would be recognized by coated pit receptor proteins. It might be anticipated that CD4 will interact with plasma membrane coated pit adaptor proteins (ROBINSON 1992). However, it is not clear whether binding will occur to the same adaptor proteins that recognise the tyrosine-containing β-turn signals, whether the signals see mutually exclusive sites on the adaptors, or whether the same signal is involved in both the coated pit and endosomal sorting of CD4. Nevertheless, it is becoming apparent that signals of this type are widely expressed in membrane proteins and, in contrast to the tyrosine containing β-turn motifs which are constitutively active, they form a class of signals that can be switched on and off at different sorting sites through the action of specific kinases and phosphatases.

### 4.9 CD4 Modulation by Gangliosides

In addition to the down-regulation induced by T cell activation and phorbol esters, treatment of cells with certain gangliosides also down-regulates CD4 expression (OFFNER et al. 1987; KAWAGUCHI et al. 1989; SAGGIORO et al. 1993). At least in part the association of gangliosides with cell surface CD4 masks the binding sites for particular anti-CD4 MAbs. However, it is also apparent that gangliosides can induce the dissociation of CD4-p56$^{lck}$ complexes and promote the internalization of cell surface CD4 (SAGGIORO et al. 1993). The mechanisms involved in these effects and their biological function (gangliosides would be complexed with serum albumin in vivo and unlikely to modulate CD4) are not understood at the present time.

## 5 CD4 Trafficking in HIV Infection and AIDS

### 5.1 HIV Entry

The human and simian immunodeficiency viruses (HIV-1, -2 and SIV) utilise CD4 as their receptor in the productive infection of T lymphocytes and monocytic cells. The binding and membrane fusion events involved in entry are widely believed to occur at the cell surface (see, for example, STEIN et al. 1987; MADDON et al. 1988; ORLOFF et al. 1991). However, given the endocytic properties of CD4 described above, viruses may also be taken into cells by endocytosis and undergo membrane fusion within endosomes (BAUER et al. 1987; PAUZA and PRICE 1988; GREWE et al. 1990; BOURINBAIAR and PHILLIPS 1991). Such a pathway for virus entry may be advantageous in certain circumstances (for review, see MARSH and PELCHEN-MATTHEWS 1993). Endocytosis may be particularly important in the infection of macrophages or dendritic cells, which are likely to have significant levels of CD4 endocytosis and which have been shown to sequester virus particles, particularly in the early asymptomatic stages of HIV infection (R.A. WEISS 1993).

## 5.2 CD4 Down-Modulation During HIV Infection

Infection of CD4 positive cells by the HIVs and SIVs leads to the loss of cell surface CD4 on the infected cells (e.g. HOXIE et al. 1986a; STEVENSON et al. 1987). As described below and in more detail elsewhere in this volume, the viruses use multiple mechanisms to control CD4 expression, indicating that cell surface CD4 expression is an important regulator of viral replication and pathogenesis. CD4 down-modulation can render infected cells resistant to super-infection by HIV, and may also facilitate the release of progeny viruses from infected cells (STEVENSON et al. 1988). In addition, by limiting the cell's ability to interact with antigen presenting cells and/or by uncoupling $p56^{lck}$, CD4 down-regulation may suppress the responsiveness of infected cells and contribute to the pathology of AIDS. Indeed it is now apparent that the expression of one of the viral proteins involved in CD4 down-regulation, Nef, confers a distinct growth advantage for HIV-1 grown in primary T cells and monocytes in culture (SPINA et al. 1994; MILLER et al. 1994), while deletion of the Nef gene from SIV generates viruses that do not appear to induce AIDS in monkeys (KESTLER et al. 1991).

## 5.3 Retention of CD4 in the Endoplasmic Reticulum

The best characterized mechanism by which HIV down-regulates CD4 expression is via retention and degradation of newly synthesised CD4 in the ER (CRISE et al. 1990; JABBAR and NAYAK 1990; CRISE and ROSE 1992a; WILLEY et al. 1992). This is particularly important late in infection, when the virus spike glycoprotein precursor gp160 is expressed at high levels. Retention is dependent on the ability of gp160 to bind to CD4 within the ER and does not occur in cells infected with viruses whose envelope proteins have a reduced affinity for CD4 (LABRANCHE et al. 1994). CD4 retention in the ER appears to be a consequence of the inefficient transport of gp160 itself (WILLEY et al. 1988; JABBAR and NAYAK 1990). In addition, the CD4/gp160 complexes formed in the ER may have conformations inappropriate for transport. The product of the viral *vpu* gene appears to augment the effects of gp160 by promoting the degradation of ER-retained CD4 (WILLEY et al. 1992). The cytoplasmic domain of CD4 (WILLEY et al. 1994), particularly a determinant between amino acids 418 and 425, appears to be important for vpu-induced CD4 degradation, but phosphorylation of the cytoplasmic serine residues or association with $p56^{lck}$ is not required (LENBURG and LANDAU 1993).

ER retrieval strategies have been used to selectively prevent delivery of HIV gp160 to the cell surface and inhibit the formation of progeny virions. Thus, sCD4 containing KDEL at the COOH-terminal allows the interaction of gp160/sCD4-KDEL complexes with cellular KDEL receptors and prevents transport of the viral envelope protein to the cell surface (BUONOCORE and ROSE 1990, 1993). Similarly, a CD4 construct with a cytoplasmic COOH-terminal ER retrieval signal, KKTC, can retain gp160 in the cell (RAJA et al. 1993), while other CD4 constructs may be used to target CD4/gp160 complexes for lysosomal degradation (LIN et al. 1993).

## 5.4 CD4 Down-Modulation by Nef

The gp160/vpu-mediated retention and degradation of CD4 in the ER occurs relatively late in infection, following synthesis of the viral structural proteins. In contrast, the HIV/SIV Nef protein is expressed early in infection and appears to induce down-regulation of cell surface CD4. The HIV *nef* gene encodes a myristoylated protein with an apparent molecular weight of 25–27 kDa, which appears to be located primarily in the cytoplasm of infected cells. Expression of the Nef protein, in the absence of other HIV proteins, will induce down-regulation of cell surface CD4, but not of human CD8, CD3 or a number of other cell surface markers (GARCIA et al. 1993). Down-regulation is believed to occur through the endocytic pathway, directing CD4 for degradation in lysosomes (AIKEN et al. 1994; ANDERSON et al. 1994; RHEE and MARSH 1994; SANFRIDSON et al. 1994). The cytoplasmic domain of CD4 is essential for Nef-induced down-regulation but, in contrast to phorbol ester-induced down-modulation, CD4 mutants lacking all three of the cytoplasmic serine residues are down-regulated by Nef (GARCIA and MILLER 1991; GARCIA et al. 1993; AIKEN et al. 1994; ANDERSON et al. 1994). Leu-413 and Leu-414 are necessary for Nef-induced CD4 modulation, but it is unclear whether these residues are required as part of an endocytosis signal or whether they are in some way involved in the interaction of CD4 with Nef (AIKEN et al. 1994). Nef expression may induce dissociation of $p56^{lck}$ from CD4 (AIKEN et al. 1994; ANDERSON et al. 1994; RHEE and MARSH 1994). In line with this the rates of CD4 endocytosis in Nef-expressing CEM cells (AIKEN et al. 1994) are similar to the bulk flow rates of CD4 uptake measured in lymphoid and nonlymphoid cells (PELCHEN-MATTHEWS et al. 1991). Precisely how Nef induces CD4 internalization is unclear. A physical interaction between CD4 and Nef has not been demonstrated in mammalian systems, although association is seen when the two proteins are co-expressed in insect cells from baculovirus vectors (HARRIS and NEIL 1994).

## 5.5 CD4 Down-Modulation by gp120

The presence of significant levels of free or antibody-complexed gp120 in the serum of HIV infected individuals and the ability of virally shed or recombinant gp120 to induce signalling via CD4 is controversial (see, for example, ORLOFF et al. 1991; KAUFMANN et al. 1992; R.A.WEISS 1993). Nevertheless, it has been shown that T cells can internalize gp120. Chronic exposure of T cell lines to recombinant gp120 can lead to dissociation of $p56^{lck}$ and, as a consequence, to limited down-regulation of CD4 (JUSZCZAK et al. 1991; CEFAI et al. 1992). After gp120 internalization and degradation, gp120-derived peptides can be presented on MHC-II molecules (LANZAVECCHIA et al. 1988) and may render CD4 positive cells susceptible to lysis by cytolytic T cells (SILICIANO et al. 1988). However, the role of these events in the destruction of CD4 positive T cells and in the pathology of AIDS remains a matter of debate (R.A.WEISS 1993).

## 6 Conclusions

CD4 is one component of the complex cell surface molecular machinery that controls the development and function of T cells. This machinery is regulated through interaction with other cells, with other proteins in the lymphocyte membrane and with cytosolic proteins. The mechanisms involved in these processes are now just beginning to be understood. Information on the interaction of CD4 with p56$^{lck}$ and with MHC-II antigens already exists. However, other interactions with, for example the TcR, components of signalling cascades, cytoskeletal elements or the protein tyrosine phosphatase CD45 (see for example Burgess et al. 1991; Guttinger et al. 1992) are less clear. These interactions may be developmentally regulated or vary according to the functional state of the cell. Information on the endocytic mechanisms that regulate CD4 cell surface expression is emerging. These trafficking properties of CD4 may also vary under these different physiologial conditions. For example, stimulation of freshly isolated T cells with anti-CD3 reagents does not induce CD4 modulation, while activated T cells readily down-regulate CD4 with identical stimuli (Rivas et al. 1988). Similarly, phorbol esters induce CD4 down-modulation on murine thymocytes, but not mature murine T cells (Wang et al. 1987; Sleckman et al. 1989; Bigby et al. 1990). There are many facets to the regulation of the cell surface expression and endocytosis of CD4, and of other T cell antigens, that remain to be unravelled. Although we do not yet fully understand all aspects of CD4 trafficking in T cell function, the presence of conserved sorting sequences in the cytoplasmic domain of the molecule suggests that these events are crucial in the regulation of T cells. Future experiments in which the effects of mutations in the specific signals are assessed in functional T cell assays will begin to establish whether or not the endocytic regulation of CD4 expression is essential for normal T cell responses.

*Acknowledgments.* We wish to thank the members of our laboratory who have contributed both to our work on CD4 and the preparation of this manuscript. In particular we thank Drs Carol Pitcher, Steve Moss, John Tite (Wellcome Foundation plc) and Jim Hoxie (University of Philadelphia) for critical comments and discussion. The authors are supported by grants from the United Kingdom Medical Research Council and the Leukaemia Research Fund.

## References

Acres BR, Conlon PJ, Mochizuki DY, Gallis B (1986) Rapid phosphorylation and modulation of the T4 antigen on cloned helper T cells induced by phorbol myristate acetate or antigen. J Biol Chem 34: 16210–16214

Aiken C, Konner J, Landau NR, Lenburg ME, Trono D (1994) Nef induces CD4 endocytosis: requirement for a critical dileucine motif in the membrane-proximal CD4 cytoplasmic domain. Cell 78: 853-864

Anderson SJ, Lenburg M, Landau NR, Garcia JV (1994) The cytoplasmic domain of CD4 is sufficient for its down-regulation from the cell surface by human immunodeficiency virus type 1 Nef. J Virol 68: 3092–3101

Arthos J, Deen KC, Chaikin MA, Fornwald JA, Sathe G, Sattentau QJ, Clapham PR, Weiss RA, McDougal JS, Pietropaolo C, Axel R, Truneh A, Maddon PJ, Sweet RW (1989). Identification of the residues in human CD4 critical for the binding of HIV. Cell 57: 469–481

Barber EK, Dasgupta JD, Schlossmman SF, Trevillyan JM Rudd CE (1989). The CD4 and CD8 antigens are coupled to a protein-tyrosine kinase (p56$^{lck}$) that phosphorylates the CD3 complex. Proc Natl Acad Sci USA 86: 3277–3281

Barclay AN, Birkeland ML, Brown MH, Beyers AD, Davis SJ, Somoza C, Williams AF (1993). The leucocyte antigen facts book. Academic, London, pp1–423

Bauer PG, Barth OM, Periera MS (1987). Endocytosis of the human immunodeficiency virus in vitro (vol 82) Membrane Institute Oswaldo Cruz, Rio de Janeiro, pp449–450

Bedinger P, Moriaty A, von Borstel II RC, Donovan NJ, Steimer KS, Littman DR (1988). Internalization of the human immunodeficiency virus does not require the cytoplasmic domain of CD4. Nature 334: 162–165

Beyers AD, Davis SJ, Cantrell DA, Izquierdo M, Williams AF (1991). Autonomous roles for the cytoplasmic domains of the CD2 and CD4 T cell surface antigens. EMBO J 10: 337–385

Beyers AD, Spruyt LL, Williams AF (1992). Molecular associations between the T-lymphocyte antigen receptor complex and the surface antigens CD2, CD4, or CD8 and CD5. Proc Natl Acad Sci USA 89: 2945–2949

Bigby M, Wang P, Fierro JF, Sy M-S (1990). Phorbol myristate acetate-induced down-modulation of CD4 is dependent on calmodulin and intracellular calcium. J Immunol 144: 3111–3116

Blue M-L, Hafler DA, Craig KA, Levine H, Schlossman SF (1988). Phosphorylation of CD4 and CD8 molecules following T cell triggering J Immunol 139: 3949–3954

Blue M-L, Daley JF, Levine H, Branton KR, Schlossman SF (1989). Regulation of CD4 and CD8 surface expression on human thymocyte subpopulations by triggering through CD2 and the CD3-T cell receptor. J Immunol 142: 374–380

Bourinbaiar AS, Philips DM (1991). Transmission of human immunodeficiency virus from monocytes to epithelia. J AIDS 4: 56–63

Brady RL, Lange G, Barclay AN (1993). Structural studies of CD4: crystal structure of domains 3 and 4 and their implication for the overall receptor structure. Biochem Soc Trans 21: 958–963

Brown D (1993) The tyrosine kinase connection: how GPI-anchored proteins activate T cells. Curr Opin Immunol 5: 349–354

Brown JH, Jardetzky TS, Gorga JC, Stern LJ, Urban RG, Strominger JL, Wiley DC (1993). Three-dimensional structure of the human class II histocompatibility antigen HLA-DRI. Nature 364: 33-39

Buonocore L, Rose JK (1990). Prevention of HIV-1 glycoprotein transport by soluble CD4 retained in the endoplasmic reticulum. Nature 345: 625–628

Buonocore L, Rose JK (1993). Blockade of human immunodeficiency virus type 1 production in CD4+ T cells by an intracellular CD4 expressed under control of the viral long terminal repeat. Proc Natl Acad Sci USA 90: 2695–2699

Burgess KE, Odysseos AD, Zalvan C, Druker BJ, Anderson P, Schlossman SF, Rudd CE (1991). Biochemical identification of a direct physical interaction between the CD4: p56$^{lck}$ and Ti(TcR)/CD3 complexes. Eur J Immunol 21: 1663–1668

Cammarota G, Schelie A, Takacs B, Doran DM, Knorr R, Bannwarth W, Guardiola J, Sinigaglia F (1992). Identification of a CD4 binding site on the β2 domain of HLA-DR molecules. Nature 356: 799–801

Cassel DL, Hoxie JA, Cooper RA (1983). Phorbol ester modulation of T-cell antigens in the Jurkat lymphoblastic leukemia cell line. Cancer Res 43: 4582–4586

Cefai D, Ferrer M, Serpente N, Idziorek T, Dautry-Varsat A, Debre P, Bismuth G (1992). Internalization of HIV glycoprotein gp120 is associated with down-modulation of membrane CD4 and p56$^{lck}$ together with impairment of T cell activation. J Immunol 149: 285–294

Clapham PR, Weiss RA, Dalgleish AG, Exley M, Whitby D, Hogg N (1987). Human immunodeficiency virus infection of monocytic and T-lymphocytic cells: receptor modulation and differentiation induced by phorbol ester. Virology 158: 44–51

Cochet C, Kashles O, Chambaz EM, Borrello I, King CR, Schlessinger. J (1988). Demonstration of epidermal growth factor-induced receptor dimerization in living cells using a chemical covalent cross-linking agent. J Biol Chem 263: 3290–3295

Cole JA, McCarthy SA, Rees MA, Sharrow SO, Singer A (1989). Cell surface comodulation of CD4 and T cell receptor by anti-CD4 monoclonal antibody. J Immunol 143: 397–402

Collins MKL, Rozengurt E (1982). Binding of phorbol esters to high affinity sites on murine fibroblastic cells elicits a mitogenic response. J Cell Physiol 112: 42–50

Collins TL, Uniyal S, Shin J, Strominger JL, Mittler RS, Burakoff SJ (1992). $p56^{lck}$ association with CD4 is required for the interaction between CD4 and the TCR/CD3 complex and for optimal antigen stimulation. J Immunol 148: 2159–2162

Crise B, Rose JK (1992a) Human immunodeficiency virus type 1 glycoprotein precursor retains a CD4-$p56^{lck}$ complex in the endoplasmic reticulum. J Virol 66: 2296–2301

Crise B, Rose JK (1992b) Identification of palmitoylation sites on CD4, the human immunodeficiency virus receptor. J Biol Chem 267: 13593–13597

Crise B, Buonocore L, Rose JK (1990). CD4 is retained in the endoplasmic reticulum by the human immunodeficiency virus type 1 glycoprotein precursor. J Virol 64: 5585–5593

Davis SJ, Ward HA, Puklavec MJ, Willis AC, Williams AF, Barclay AN (1990). High level expression in chinese hamster ovary cells of soluble forms of CD4 T lymphocyte glycoprotein including glycosylation variants. J Biol Chem 265: 10410–10418

Dietrich J, Hou X, Wegener A-M, Geisler C (1994). CD3γ contains a phosphoserine-dependent dileucine motif involved in down-regulation of the T cell receptor. EMBO J 13: 2156–2166

Dwek RA, Ashford DA, Edge CJ, Parekh RB, Rademacher TW, Wing DR, Barclay AN, Davis SJ, Williams AF (1993). Glycosylation of CD4 and Thy-1. Philos Trans R Soc Lond(Biol) 342: 43–50

Fleury S, Lamarre D, Meloche S, Ryu SE, Cantin C, Hendrickson WA, Sekaly RP (1991). Mutational analysis of the interaction between CD4 and class II MHC: class II antigens contact CD4 on a surface opposite the gp120-binding site. Cell 66: 1037–1049

Garcia JV, Miller AD (1991). Serine phosphorylation-independent downregulation of cell-surface CD4 by nef. Nature 350: 508–511

Garcia JV, Alfano J, Miller AD (1993). The negative effect of human immunodeficiency virus type 1 nef on cell surface CD4 expression is not species specific and requires the cytoplasmic domain of CD4. J Virol 67: 1511–1516

Glaichenhaus N, Shastri N, Littman DR, Turner JM (1991) Requirement for association of $p56^{lck}$ with CD4 in antigen-specific signal transduction in T cells. Cell 64: 511–520

Grewe C, Beck A, Gelderblom HR (1990). HIV: early virus-cell interactions. J AIDS 3: 965–974

Griffiths G, Back R, Marsh M (1989). A quantitative analysis of the endocytic pathway in Baby Hamster Kidney cells. J Cell Biol 109: 2703-2720

Guttinger M, Gassman M, Amrein KE, Burn P (1992). CD45 phosphotyrosine phosphatase and $p56^{lck}$ protein tyrosine kinase: a functional complex crucial in T cell signal transduction. Int Immunol 4: 1325–1330

Hansen S, Sandvig K, van Deurs B (1992). Internalisation efficiency of the transferrin receptor. Exp Cell Res 199: 19–28

Harris MPG, Neil JC (1994). Myristoylation-dependent binding of HIV-1 Nef to CD4. J Mol Biol 241: 136–142

Hatakeyama M, Kono T, Kobayashi N, Kawahara A, Levin SD, Perlmutter RM, Taniguchi T (1991). Interaction of the IL-2 receptor with the src-family kinase $p56^{lck}$: identification of novel intermolecular association. Science 252: 1523–1528

Helenius A (1994). How N-linked oligosaccharides affect glycoprotein folding in the endoplasmic reticulum. Mol Biol Cell 5: 253-265

Hoxie JA, Alpers JD, Rackowski JL, Huebner K, Haggarty BS, Cedarbaum AJ, Reed JC (1986a). Alterations in T4 (CD4) protein and mRNA synthesis in cells infected with HIV. Science 234: 1123–1127

Hoxie JA, Matthews DM, Callahan KJ, Cassel DL, Cooper RA (1986b). Transient modulation and internalization of T4 antigen induced by phorbol esters. J Immunol 137: 1194–1201

Hoxie JA, Rackowski JL, Haggarty BS, Gaulton GN (1988). T4 Endocytosis and phosphorylation induced by phorbol esters but not by mitogen or HIV infection. J Immunol 140: 786–795

Hurley TR, Luo K, Sefton BM (1989). Activators of protein kinase induce dissociation of CD4, but not CD8, from $p56^{lck}$ Science 245: 407–409

Jabbar MA, Nayak DP (1990). Intracellular interaction of human immunodeficiency virus type 1 (ARV-2) envelope glycoprotein gp160 with CD4 blocks the movement and maturation of CD4 to the plasma membrane. J Virol 64: 6297–6304

Janeway CAJ (1989). The role of CD4 in T-cell activation: accessory molecule or co-receptor? Immunol Today 10: 234–238

Juszczak RJ, Turchin H, Truneh A, Culp J, Kassis S (1991). Effect of human immunodeficiency virus gp120 glycoprotein on the association of the protein tyrosine kinase $p56^{lck}$ with CD4 in human T lymphocytes. J Biol Chem 266: 11176–11183

Kammer GM, Boehm CA, Rudolph SA, Schultz LA (1988). Mobility of the human T lymphocyte surface molecules CD3, CD4 and CD8: regulation by a cAMP-dependent pathway. Proc Natl Acad Sci USA 85: 792–796

Kaufmann R, Laroche D, Buchner K, Hucho F, Rudd C, Lindschau C, Ludwig P, Höer A, Oberdisse E, Kopp J, Körner I-J, Repke H (1992). The HIV–1 surface protein gp120 has no effect on transmembrane signal transduction in T cells. J AIDS 15: 760–770

Kawaguchi T, Nakakuma H, Kagimoto T, Shirono K, Horikawa K, Hidaka M, Iwamori M, Nagai Y, Takatsuki K (1989). Characteristic mode of action of gangliosides in selective modulation of CD4 on human T lymphocytes. Biochem Biophys Res Commun 158: 1050–1059

Keller G-A, Siegel MW, Caras IW (1992). Endocytosis of glycosphingolipid-anchored and transmembrane forms of CD4 by different endocytic pathways. EMBO J 11: 863–874

Kennelly PJ, Krebs EG (1991). Consensus sequences as substrate specificity determinants for protein kinases and protein phosphatases. J Biol Chem 266: 15555–15558

Kestler HW, Ringler DJ, Mori K, Panicali DL, Sehgal PK, Daniel MD, Desrosiers RD (1991). Importance of the nef gene for maintenance of high virus loads and for development of AIDS. Cell 65: 651–662

Kim Y-J, Pollok KE, Zhou Z, Shaw A, Bohlen JB, Fraser M, Kwon BS (1993). Novel T cell antigen 4–1BB associates with the protein tyrosine kinase $p56^{lck}$. J Immunol 151: 1255–1262

Knight SC, Macatonia SE, Patterson S (1990). HIV 1 infection of dendritic cells. Int Rev Immunol 6: 163–175

König R, Huang L-Y, Germain R (1992). MHC class II interaction with CD4 mediated by a region analogous to the MHC class 1 binding site for CD8. Nature 356: 796–798

Kwong PD, Ryu S-E, Hendrickson WA, Axel R, Sweet RM, Folena-Wasserman G, Hensley P, Sweet RW (1990). Molecular characteristics of recombinant human CD4 as deduced from polymorphic crystals. Proc Natl Acad Sci USA 87: 6423-6427

LaBranche CC, Sauter MM, Haggarty BS, Vance PJ, Romano J, Hart TK, Bugelski PJ, Hoxie JA (1994). Biological, molecular and structural analysis of a cytopathic variant from a molecularly cloned simian immunodeficiency virus. J Virol 68: 5509–5522

Lanzavecchia A, Roosnek E, Gregory T, Berman P, Abrignani S (1988). T cells can present antigens such as HIV gp120 targeted to their own surface molecules. Nature 334: 530–532

Lenburg ME, Landau NR (1993). Vpu-induced degradation of CD4: requirement for specific amino acid residues in the cytoplasmic domain of CD4. J Virol 67: 7238–7245

Letourneur F, Klausner RD (1992). A novel di-leucine motif and a tyrosine-based motif independently mediate lysosomal targetting and endocytosis of CD3 chains. Cell 69: 1143-1157

Lin X, Dashti A, Schinazi RF, Tang J (1993). Intracellular diversion of glycoprotein gp160 of human immunodeficiency virus to lysosomes as a strategy of AIDS gene therapy. FASEB J. 7: 1070–1080

Littman DR, Maddon PJ, Axel R (1988). Corrected CD4 sequence. Cell 55: 541

Louie RR, King CS, MaCauley A, Marth JD, Perlmutter RM, Eckhart W, Cooper JA (1988). $p56^{lck}$ protein-tyrosine kinase is cytoskeletal and does not bind to polyomavirus middle T antigen. J Virol 62: 4673-4679

Maddon PJ, Littman DR, Godfrey M, Maddon DE, Chess L, Axel R (1985). The isolation and nucleotide sequence of a cDNA encoding the T cell surface protein T4: A new member of the immunoglobulin gene family. Cell 42: 93–104

Maddon PJ, McDougal JS, Clapham PR, Dalgleish AG, Jamal S, Weiss RA, Axel R (1988). HIV infection does not require endocytosis of its receptor, CD4. Cell 54: 865–874

Marchildon GA, Casnellie JE, Walsh KA, Krebs EG (1984). Covalently bound myristate in a lymphoma tyrosine protein kinase. Proc Natl Acad Sci USA 81: 7679–7682

Marsh M, Helenius A (1980). Adsorptive endocytosis of Semliki Forest virus. J Mol Biol 142: 439–454

Marsh M, Pelchen-Matthews A (1993). Animal virus entry into cells. Rev Med Virol 3: 173–185

Marsh M, Reid P, Parsons I, Hermon C, Pelchen-Matthews A (1992). Morphological analysis of the regulation of CD4 endocytosis by $p56^{lck}$. Biochem Soc Trans 20: 719–724

Marth JD, Peet R, Krebs EG, Perlmutter RM (1985). A lymphocyte-specific protein tyrosine kinase gene is rearranged and overexpressed in the murine T cell lymphoma LSTRA. Cell 43: 393-404

Mellman I, Fuchs R, Helenius A (1986). Acidification of the endocytic and exocytic pathways. Annu Rev Biochem 55: 663-700

Miller K, Shipman M, Trowbridge IS, Hopkins CR (1991). Transferrin receptors promote the formation of clathrin lattices. Cell 65: 621–632

Miller MD, Warmerdam MT, Gaston I, Greene WC, Feinberg MB (1994) The human immunodeficiency virus-1 *nef* gene product. a positive factor for viral infection and replication in primary lymphocytes and macrophages. J Exp Med 179: 101–113

Mittler RS, Hoffman MK (1989). Synergism between gp120 and gp120-specific antibody in blocking human T cell activation. Science 245: 1380–1382

Newell MK, Haughn LJ, Maroun CR, Julius MH (1990). Death of mature T cells by separate ligation of CD4 and the T-cell receptor for antigen. Nature 347: 286–289

Offner H, Thieme T, Vandenbark AA (1987). Gangliosides induce selective modulation of CD4 from helper T lymphocytes. J Immunol 139: 3295–3305

Orloff GM, Orloff SL, Kennedy MS, Maddon PJ, McDougal JS (1991). Penetration of CD4 T cells by HIV 1. The CD4 receptor does not internalize with HIV, and CD4 related signal transduction events are not required for entry. J Immunol 146: 2578–2587

Paige LA, Nadler MJS, Harrison ML, Cassady JM, Geahlen RL (1993). Reversible palmitoylation of the protein-tyrosine kinase $p56^{lck}$. J Biol Chem 268: 8669–8674

Pal R, Nair BC, Hoke GM, Sarngadharan MG, Edidin M (1991). Lateral diffusion of CD4 on the surface of a human neoplastic T-cell line probed with a fluorescent derivative of the envelope glycoprotein (gp120) of human immunodeficiency virus type 1 (HIV-1). J Cell Physiol 147: 326–332

Pauza CD, Price TM (1988). Human immunodeficiency virus infection of T cells and monocytes proceeds via receptor-mediated endocytosis. J Cell Biol 107: 959–968

Pelchen-Matthews A, Armes JE, Marsh M (1989). Internalization and recycling of CD4 transfected into HeLa and NIH-3T3 cells. EMBO J 8: 3641–3649

Pelchen-Matthews A, Armes JE, Griffiths G, Marsh M (1991). Differential endocytosis of CD4 in lymphocytic and non-lymphocytic cells. J Exp Med 173: 575–587

Pelchen-Matthews A, Boulet I, Littman D, Fagard R, Marsh M (1992). The protein tyrosine kinase $p56^{lck}$ inhibits CD4 endocytosis by preventing entry of CD4 into coated pits. J Cell Biol 117: 279–290

Pelchen-Matthews A, Parsons IJ, Marsh M (1993). Phorbol ester-induced down-regulation of CD4 is a multi-step process involving dissociation from $p56^{lck}$, increased association with clathrin-coated pits and altered endosomal sorting. J Exp Med 178: 1209–1222

Podack ER, Kupfer A (1991). T-cell effector functions: mechanisms for delivery of cytotoxicity and help. Annu Rev Cell Biol 7: 479–504

Raab M, Yamamoto M, Rudd CE (1994) The T-cell antigen CD5 acts as a receptor and substrate for the protein-tyrosine kinase $p56^{lck}$. Mol Cell Biol 14: 2862–2870

Raja NU, Vincent MJ, Jabbar MA (1993). Analysis of endoproteolytic cleavage and intracellular transport of human immunodeficiency virus type 1 envelope glycoproteins using mutant CD4 molecules bearing the transmembrane endoplasmic reticulum retention signal. J Gen Virol 74: 2085–2097

Resh MD (1990). Membrane interactions of $pp60^{v\text{-}src}$: a model for myristylated tyrosine protein kinases. Oncogene 5: 1437–1444

Rhee SS, Marsh JW (1994). Human immunodeficiency virus type 1 Nef-induced down-modulation of CD4 is due to rapid internalization and degradation of surface CD4. J Virol 68: 5156–5163

Rivas A, Takada S, Koide J, Sonderstrup-McDevitt G, Engleman EG (1988). CD4 molecules are associated with the antigen receptor complex on activated but not resting T cells. J Immunol 140: 2912–2918

Robey E, Axel R (1990). CD4: collaborator in immune recognition and HIV infection. Cell 60: 697–700

Robinson M (1992). Adaptins. Trends Cell Biol 2: 293–297

Rosenstein Y, Burakoff SJ, Herrmann SH (1990). HIV-gp120 can block CD4-class II MHC-mediated adhesion. J Immunol 144: 526–531

Rudd CE (1990). CD4, CD8 and the TCR-CD3 complex: a novel class of protein-tyrosine kinase receptor. Immunol Today 11: 400-406

Rudd CE, Trevillyan JM, Dasgupta JD, Wong LL, Schlossman SF (1988). The CD4 receptor is complexed in detergent lysates to a protein-tyrosine kinase (pp58) from human T lymphocytes. Proc Natl Acad Sci USA 85: 5190–5194

Ruegg CL, Rajasekar S, Stein BS, Engleman EG (1992). Degradation of CD4 following phorbol-induced internalization in human T lymphocytes. Evidence for distinct endocytic routing of CD4 and CD3. J Biol Chem 267: 18837–18843

Ryu S-E, Kwong PD, Truneh A, Porter TG, Arthos J, Rosenberg M, Dai X, Xuong N, Axel R, Sweet RW, Hendrickson WA (1990). Crystal structure of an HIV-binding fragment of human CD4. Nature 348: 419–426

Saggioro D, Sorio C, Calderazzo F, Callegaro L, Panozzo M, Berton G, Chieco-Bianchi L (1993). Mechanism of action of the monosialoganglioside GM1 as a modulator of CD4 expression. J Immunol 268: 1368–1375

Saizawa K, Rojo J, Janeway CAJ (1987). Evidence for a physical association of CD4 and the CD3: $\alpha$: $\beta$ T-cell receptor. Nature 328: 260–263

Sako Y, Kusumi A (1994). Compartmentalized structure of the plasma membrane for receptor movements as revealed by nanometer-level motion analysis. J Cell Biol 125: 1251–1264

Sanfridson A, Cullen BR, Doyle C (1994). The simian immunodeficiency virus nef protein promotes degradation of CD4 in human T cells. J Biol Chem 269: 3917–3920

Sattentau QJ, Weiss RA (1988). The CD4 antigen: physiological ligand and HIV receptor. Cell 52: 631–633

Sawada S, Scarborough JD, Killeen N, Littman DR (1994). A lineage specific transcriptional silencer regulates CD4 gene expression during T lymphocyte development. Cell 77: 917–929

Schmid SL (1992). The mechanism of receptor-mediated endocytosis: more questions than answers. Bioessays 14: 589–596

Shaw AS, Amrein KE, Hammond C, Stern DF, Sefton BM, Rose JK (1989) The Ick tyrosine protein kinase interacts with the cytoplasmic tail of the CD4 glycoprotein through its unique amino-terminal domain. Cell 59: 627–636

Shaw AS, Chalupny J, Whitney JA, Hammond C, Amrein KE, Kavathas P, Sefton BM, Rose JK (1990). Short related sequences in the cytoplasmic domains of CD4 and CD8 mediate binding to the amino-terminal domain of the $p56^{lck}$ tyrosine protein kinase. Mol Cell Biol 10: 1853–1862

Shenoy-Scaria AM, Dietzen DJ, Kwong J, Link DC, Lublin DM (1994) Cysteine[3] of Src family protein tyrosine kinase determines palmitoylation and localization in caveolae. J Cell Biol 126: 353–363

Shin J, Doyle C, Yang Z, Kappes D, Strominger JL (1990). Structural features of the cytoplasmic region of CD4 required for internalization. EMBO J 9: 425–434

Shin J, Dunbrack RI, Lee S, Strominger JL (1991). Phosphorylation-dependent down-modulation of CD4 requires a specific structure within the cytoplasmic domain of CD4. J Biol Chem 266: 10658–10665

Siliciano RF, Lawton T, Knall C, Karr RW, Berman P, Gergory T, Reinherz EL (1988). Analysis of host-virus interactions in AIDS with anti-gp120 T cell clones: effect of HIV sequence variation and mechanism of $CD4^+$ cell depletion. Cell 54: 561–575

Sleckman BP, Bigby M, Greenstein JL, Burakoff SJ, Sy M-S (1989). Requirements for modulation of the CD4 molecule in response to phorbol myristate acetate. J Immunol 142: 1457–1462

Sleckman BP, Shin J, Igras VE, Collins TL, Strominger JL, Burakoff SJ (1992). Disruption of the CD4-$p56^{lck}$ complex is required for rapid internalization of CD4. Proc Natl Acad Sci USA 89: 7566–7570

Spina CA, Kwoh TJ, Chowers MY, Guatelli JC, Richman DD (1994) The importance of nef in the induction of human immunodeficiency virus type 1 replication from primary quiescent CD4 lymphocytes. J Exp Med 179: 115–123

Stein BS, Gowda SD, Lifson JD, Penhallow RC, Bensch KG, Engleman EG (1987). pH-Independent HIV entry into CD4-positive T cells via virus envelope fusion to the plasma membrane. Cell 49: 659–668

Stevenson M, Zhang X, Volsky DJ (1987). Downregulation of cell surface molecules during noncytopathic infection of T cells with human immunodeficiency virus. J Virol 61: 3741–3748

Stevenson M, Meier C, Mann AM, Chapman N, Wasiak A (1988). Envelope glycoprotein of HIV induces interference and cytolysis resistance in $CD4^+$ cells: mechanism for persistence in AIDS. Cell 53: 483–496

Suzuki H, Zelphati O, Hildebrand G, Leserman L (1991). CD4 and CD7 molecules as targets for drug delivery from antibody bearing liposomes. Exp Cell Res 193: 112–119

Taieb J, Vitte-Mony I, Auffredou MT, Dorseuil O, Gacon G, Bertoglio J, Vazquez A (1993) Regulation of $p56^{lck}$ kinase expression and control of DNA synthesis in activated human B lymphocytes. J Biol Chem 268: 9169–9171

Tite JP, Sloan A, Janeway CA (1986). The role of L3T4 in T cell activation: L3T4 may be both an Ia-binding protein and a receptor that transduces a negative signal. J Mol Cell Immunol 2: 179–190

Trowbridge IS, Collawn JF, Hopkins CR (1993). Signal-dependent membrane protein trafficking in the endocytic pathway. Annu Rev Cell Biol 9: 129–161

Truner JM, Brodsky MH, Irving BA, Levin SD, Perlmutter RM, Littman DR (1990). Interaction of the unique N-terminal region of tyrosine kinase $p56^{lck}$ with cytoplasmic domains of CD4 and CD8 is mediated by cysteine motifs. Cell 60: 755–765

van Deurs B, Peterson OW, Olsens S, Sandvig K (1989). The ways of endocytosis. Int Rev Cytol 117: 131–177

Vaux D (1992). The structure of an endocytosis signal. Trends Cell Biol 2: 189–192

Veillette A, Bookman MA, Horak EM, Samelson LE, Bolen JB (1989). Signal transduction through the CD4 receptor involves the activation of the internal membrane tyrosine-kinase $p56^{lck}$. Nature 338: 257–259

Voronova AF, Sefton BM (1986). Expression of a new tyrosine protein kinase is stimulated by retrovirus promoter insertion. Nature 319: 682–685

Wang J, Yan Y, Garrett TPJ, Liu J, Rodgers DW, Garlick RL, Tarr GE, Husain Y, Reinherz EL, Harrison SC (1990). Atomic structure of a fragment of human CD4 containing two immunoglobulin-like domains. Nature 348: 411–418

Wang PTH, Bigby M, Sy M-S (1987). Selective down modulation of L3T4 molecules on murine thymocytes by the tumor promoter, phorbol 12-myristate 13-acetate. J Immunol 139: 2157–2165

Watts C, Marsh M (1992). Endocytosis: what goes in and how ? J Cell Sci 103: 1–8

Weiss A (1993). T cell antigen receptor signal transduction: a tale of tails and cytoplasmic protein-tyrosine kinases. Cell 73: 209–212

Weiss RA (1993). How does HIV cause AIDS? Science 260: 1273-1279

Weyand CM, Goronzy J, Fathman CG (1987). Modulation of CD4 by antigenic activation. J Immunol 138: 1351–1354

Willey RL, Bonifacino JS, Potts BJ, Martin MA, Klausner RD (1988). Biosynthsis, cleavage, and degradation of the human immunodeficiency virus 1 envelope glycoprotein gp160. Proc Natl Acad Sci USA 85: 9580–9584

Willey RL, Maldarelli, F, Martin MA, Strebel K (1992). Human immunodeficiency virus type 1 vpu protein regulates the formation of intracellular gp160-CD4 complexes. J Virol 66: 226–234

Willey RL, Buckler-White A, Strebel K (1994). Sequences present in the cytoplasmic domain of CD4 are necessary and sufficient to confer sensitivity to the human immunodeficiency virus type 1 Vpu protein. J Virol 68: 1207–1212

Xu H, Littman DR (1993). A kinase-independent function of Lck in potentiating antigen-specific T cell activation. Cell 74: 633-643

Yoshida H, Koga Y, Nakamura K, Kimura G, Nomoto K (1992). A lymphocyte-specific protein tyrosine kinase, $p56^{lck}$, regulates the PMA-induced internalization of CD4. Biochim Biophys Acta 1137: 321–330

Zhang F, Lee GM, Jacobson K (1993) Protein lateral mobility as a reflection of membrane microstructure. Bioessays 15: 579–588

# The Receptor for HIV: Dissection of CD4 and Studies on Putative Accessory Factors

W. James[1], R.A. Weiss[2], and J.H.M. Simon[3]



## 1 Introduction

Soon after the identification of HIV as the causative agent of AIDS, it became clear that CD4 was its primary cellular receptor. The clinical observation that

[1]Sir William Dunn School of Pathology, University of Oxford, South Parks Road, Oxford, OX1 3RE, UK
[2]Chester Beatty Laboratories, Institute of Cancer Research, 237 Fulham Road, London, SW3 6JB, UK
[3]Howard Hughes Medical Institute, University of Pennsylvania School of Medicine, Clinical Research Building, 422 Curie Boulevard, Philadelphia, PA 19104, USA

AIDS was characterized by a loss of the T helper/inducer subset of lymphocytes ($T_h$) prompted the studies which showed that HIV replicated in vitro only in cells expressing the $T_h$ marker, CD4 (KLATZMANN et al. 1984a), and that antibodies to CD4 specifically inhibited infection and syncytium formation (DALGLEISH et al. 1984; KLATZMANN et al. 1984b). The receptor function of CD4 was conclusively demonstrated when its expression was shown to confer HIV susceptibility on previously insusceptible cells (MADDON et al. 1986). Finally, when recombinant soluble CD4 was mixed with virus or virus-infected cells, infection and syncytium formation were inhibited, presumably because of competition with the cell surface receptor (SMITH et al. 1987; FISHER et al. 1988; HUSSEY et al. 1988; TRAUNECKER et al. 1988).

Two major questions remained. First, what regions of CD4 are required for cellular infection by HIV and what are their precise functions in that process? Second, what other cellular molecules, if any, are required for efficient HIV entry? In this chapter, we review the experiments that have largely answered the first question and have begun to shed light on the second.

## 2 The Dissection of CD4

### 2.1 Identification of the gp120 Binding Site

It was several years until the atomic structure of human CD4 (hCD4) domains 1 and 2 (RYU et al. 1990; WANG et al. 1990) and rat CD4 domains 3 and 4 (BRADY et al. 1993) became available. In the mean time, a combination of monoclonal antibodies (mAbs) and mutagenesis was employed to map the epitope structure of CD4 and to define which regions were involved in HIV entry.

Early experiments showed that not all anti-CD4 mAbs inhibited HIV infection and that those which did tended to cluster in domains 1 and 2 (McDOUGAL et al. 1986; SATTENTAU et al. 1986). Using synthetic peptides and mutants of CD4, the epitope of a cluster of mAbs that inhibited the binding of HIV gp120 to CD4, typified by OKT4A, had been mapped to residues 32–57 of domain 1, which was correctly thought to comprise the C'C" region (JAMESON et al. 1988; MIZUKAMI et al. 1988).

Studies using truncated forms of CD4 showed that the gp120 binding activity lay in the $NH_2$-terminal half of the molecule (RICHARDSON et al. 1988; TRAUNECKER et al. 1988). Mutagenesis of these two domains, either by replacement of hCD4 residues with their nonconservative variant homologues in murine CD4 or by insertion of an Arg-Ser dipeptide at selected sites, confirmed the work from antibody studies showing that the gp120 binding site was contained within the first domain (CLAYTON et al. 1988; MIZUKAMI et al. 1988). The mutations that abrogated binding were concentrated in the C'C" region but some were found at more distant sites, including within domain 2. A secreted form of domain 1 was shown to have full gp120 binding properties and a set of mutants with alterations in this domain again implicated residues between positions 42 and 55 in gp120 binding (ARTHOS et al. 1989). A more comprehensive set of mutations replacing hCD4 residues with nonconserved murine CD4 residues emphasized the importance of amino acids 41–59, but again there appeared to be contributions from a few more distant sites (BRODSKY et al. 1990). In a comprehensive study of domain 1, in which each of the 64 residues was mutated to alanine, the picture was confirmed, with significant residues clustering between positions 39 and 52 and isolated amino acids at Gln-64, Lys-29 and Glu-85 (ASHKENAZI et al. 1990). Impairment in gp120 binding was observed in a fortuitous deamidation of Asn-52 to Asp in a preparation of soluble CD4 (TESHIMA et al. 1991).

The problem with such studies is distinguishing between mutations within the contact zone and those acting at a distance that disrupt the precise folding of the domain. In most studies, a panel of conformation-sensitive mAbs was used as a control for gross disruptive effects. However, it is hard to be sure that loss of a particular epitope reflects a change at the antibody binding site per se rather than long-range effects or that all significant conformational changes can be detected in this way. Once the atomic structure was published, it became easier to identify which residues were involved in interchain H-bonding and which might be free to be mutated without major structural deformation.

By using a phage display method to screen for mutations between residues 38 and 62 that affected gp120 binding, the region between 41 and 51 was shown again to be of prime importance (TSUI et al. 1992). These mutations lie in the exposed C" ridge and the turns preceding and following it, while other mutations (some of which increased gp120 binding affinity) lie in the underlying C' and D strands, probably exerting their effects indirectly by altering the conformation of the contact residues in C".

By reference to the atomic structure of CD4 and examining the detailed binding kinetics of mutants to gp120, two studies concluded that the gp120 binding site was the projecting phenylalanine residue at position 43 (top of C"), flanked by positively charged residues brought together by the folding of the β strands (Fig. 1). Lys-46 and Arg-59 were of clear importance but the roles of Lys-29 and Lys-35 were more controversial (CHOE and SODROSKI 1992; MOEBIUS et al. 1992). Clearly, any mutation outside the contact site that affects the correct alignment of these strands might abolish gp120 binding.

A competent gp120 binding site was reconstructed in rat CD4, which shares only 50% amino acid homology with human CD4 and has no shared epitopes, by replacement of just the C'C"D region (residues 33–62; Fig. 1) by the equivalent region from the human homologue (SCHOCKMEL et al. 1992). This region is more extensive than the probable contact site but shorter substitutions were not fully wild-type in their ability to bind gp120—probably because of the need to maintain interstrand bonding necessary for correct presentation of the binding site.

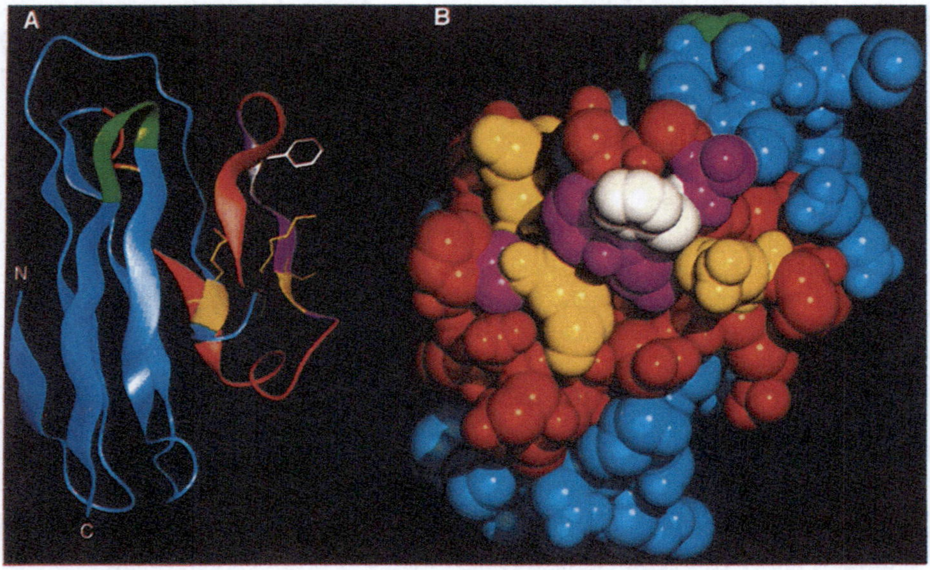


**Fig. 1 A, B.** The HIV receptor. **A** A ribbon diagram of domain 1 of human CD4 from the standard point of view The $NH_2$-terminus of strand A and the COOH-terminus of strand G are marked. **B** A space-filling model of domain 1 of human CD4 from the point of view of a monomer of gp120 approaching the contact site. In both panels, residues are coloured similarly. Residues 86–89 (F-G turn, equivalent to CDR3 of Ig variable domains) are shown in *green*. Residues 33–62 (C'-C"-D, equivalent to CDR2) are shown in *red*, with the exception of the following: Residues Lys-29, Lys-35, Lys-46 and Arg-59 are shown, with their side chains, in *yellow*. Residue Phe-43, with its side chain, is shown in *off-white*. The remaining residues of the C" ridge (Ser-42, Leu-44, Thr-45, Gly-47) are shown in *magenta*. All other residues are shown in *blue*

## 2.2 The Pursuit of Anti-idiotype gp120 Binding Site Mimics

It was possible that some antibodies might have made contact with CD4 in the same manner as gp120. These antibodies would mimic gp120 and could be used to stimulate the production of neutralizing anti-idiotype antibodies (McDougal et al. 1986) but they would also be expected to not bind mutants in the gp120 binding site. One early paper reported that polyclonal anti-idiotypes to the CD4 mAb, Leu3a did indeed neutralize a panel of HIV strains (Dalgleish et al. 1987). Although a careful study of four highly selected mAbs failed to reveal neutralizing anti-idiotypes (Healey et al. 1991), another study appeared to show the production of neutralizing, gp120 binding anti-idiotypes to mAbs directed at both the CDR2 and CDR3 regions of CD4 (Sutor et al. 1992). The idea was put to rest with a study of 225 anti-CD4 mAbs whose fine specificity and affinity were compared with those of gp120. None matched the receptor-ligand interaction and this, together with the poor historic performance of the approach, suggested that the method was unlikely to be successful and that positive results should be treated sceptically (Davis et al.1992).

## 2.3 The Role of Sites Outside CDR2 in HIV Receptor Function

### 2.3.1 CDR3

A variety of studies suggested that sites other than those believed to be in the gp120 contact zone were specifically involved in postattachment stages of virus entry. For example, a cluster of mAbs that recognized the FG turn of domain1, homologous to the CDR3 loop of Ig V domains, were shown to inhibit HIV infection without interfering with gp120 binding (TRUNEH et al. 1991). The CDR3 region is largely obscured by the B-C loop (CDR1) when viewed from the gp120 binding site (Fig. 1B). The proposal was that this region specifically interacted with regions of HIV gp120 or gp41 that were exposed following the initial interaction with the high-affinity site within C'C" and that this interaction ws necessary for subsequent steps, including virus envelope-plasma membrane fusion.

The importance of this region was reinforced by studies in which synthetic peptides corresponding to portions of domain 1 of CD4 were screened for their ability to bind gp120 or inhibit the binding of gp120 to CD4. Peptides from the CDR2-like region of CD4 occasionally had activity in these assays (JAMESON et al. 1988; CEREDA et al. 1991) but often did not (KALYANARAMAN et al. 1990, OHKI et al. 1994). The most potent peptides were derived from the F-G turn or CDR3-like region (LIFSON et al. 1988; NARA et al. 1989; KALYANARAMAN et al. 1990; SHAPIRANAHOR et al. 1990; LIFSON et al. 1991; RAUSCH et al. 1990, 1992; MESHCHERYAKOVA et al. 1993; LASARTE et al. 1994; OHKI et al. 1994). It was found that these peptides could mimic CD4 by promoting the dissociation of gp120 from gp41 (BERGER et al. 1991; see Sect. 2.5.2).

This consensus correlated with the antibody studies and with studies on chimpanzee CD4, in which the substitution of Glu for Gly at residue 87 was reported to abolish the ability to support HIV-mediated syncytia (CAMERINI and SEED 1990; LIFSON et al. 1991). Further evidence suggested that the V3 region of gp120 (the principal neutralizing domain) directly interacted with CDR3 to promote virus-cell fusion (AUTIERO et al. 1991; EIDEN and LIFSON 1992; MESHCHERYAKOVA et al. 1993).

However, this picture has been discredited. First, the CDR3 peptides were found to inhibit the binding of a CDR2-directed mAb and syncytium formation by the unrelated and CD4-independent virus HTLV-1. In addition, their activity was found to depend crucially on their derivitization with aromatic groups (REPKE et al. 1992). Second, a chimeric CD4 molecule comprising rat CD4, in which the C'C"D region of D1 was replaced by that of human CD4, was found to act as an efficient receptor for HIV despite having an entirely different CDR3 sequence from human CD4 (SIMON et al. 1993). Finally, a detailed study of CDR3-region mutants revealed that a large panel of mutations in the CDR3 region had no effect on syncytium formation unless they also had gross effects on the structure of CD4 which led to disruption of gp120 binding (BRODER and BERGER 1993).

More recently, the mechanism by which anti-CDR3 antibodies inhibit HIV infection has become clear, (see Sect. 2.4).

### 2.3.2 Domains 2, 3 and 4

Studies using anti-CD4 mAbs have implicated regions of CD4 other than the gp120 binding site in the post binding events that lead to plasma membrane-virus envelope fusion. A mAb against domain 2 was shown to block infection any syncytium formation without preventing binding of soluble CD4 to virions. However, conformational changes in the viral envelope glycoproteins that lead to exposure of cryptic epitopes on gp41 and its dissociation from gp120 were inhibited by this antibody, suggesting further roles for CD4 during infection (BURKLY et al. 1992; MOORE et al. 1992). An antibody against domain 3, while not inhibiting virion attachment, was inhibitory to infection and syncytium formation (HEALEY et al. 1990). The epitope was thought to lie adjacent to the residues that connect domains 2 and 3, in which a degree of flexibility might lie.

Consequently, it was proposed that flexion about this "hinge" might be an essential post attachment event. Preliminary results with a first-round of mutants in which the hinge was partially deleted imply that flexibility in this region is important for HIV entry (Q.J. Sattentau, personal communication). Finally, a mAb to domains 3 and 4 was shown to inhibit infection without affecting syncytium formation or binding (HASUNUMA et al. 1992).

The evidence for the importance of these regions is less clear from studies of CD4 chimeras. A glycoprotein consisting of the $NH_2$-terminal two domains of CD4 and the hinge, transmembrane and cytoplasmic domains of CD8 was a functional receptor for HIV (POULIN et al. 1991). Although the rates of HIV replication and syncytium formation were lower than with wild-type CD4, this was also true to a somewhat lesser extent with CD4 mutants truncated only in the cytoplasmic tail. It was later shown that syncytium formation, believed to be blocked in cells expressing the CD4: CD8 chimeras (POULIN et al. 1991), was simply delayed (GOLDING et al. 1993). Interpretation of these results is complicated by the fact that the portions of CD8 present in the chimeras are those that promote its dimerization. Given the multisubunit interactions between gp120 and CD4 now believed to be involved in HIV entry (see Sect. 2.4), inappropriate orientation on CD4 domains 1 promoted by CD8-mediated dimerization might have effects on the rate of infection.

The rat-human CD4 chimera humanized only in the gp120 binding region of domain 1 was an efficient HIV receptor (SIMON et al. 1993), suggesting that, if regions of domains 2, 3 and 4 are involved in HIV entry, their role must be structural or dynamic and not involve specific intermolecular recognition.

## 2.4 Antibody Studies Reveal Multimeric gp120-CD4 Interactions

An immunoglobulin molecule is approximately three times more bulky than CD4 itself. Consequently, the question remains whether the inhibition of HIV replication by an antibody really reveals an essential and specific function of the region to which it is bound or whether the antibody interferes sterically with events occurring elsewhere.

Two studies on the concentration and temperature dependence of soluble CD4-virion binding, gp120 dissociation and inhibition of HIV entry showed that the processes were highly cooperative and suggested that the interaction of multiple gp120 molecules within the virion envelope glycoprotein oligomers with CD4 was required for infection (LAYNE et al. 1990; MOORE et al. 1991). Experiments probing the mobility of CD4 expressed at the cell surface also suggested that CD4 frequently clustered (PAL et al. 1991). Although it has been proposed that the CDR3 region is involved in dimerization (LANGEDIJK et al. 1993), this is contentious (see Brady and Barclay, this volume).

It has recently emerged that, while CDR3-directed mAbs do not inhibit the interaction of monomeric gp120 with soluble CD4 (sCD4), they partially inhibit the binding of sCD4 to virions (MOORE 1993). This observation can be explained as follows. The interaction between one copy of gp120 within a virus envelope spike glycoprotein oligomer and a molecule of CD4 is not inhibited by the presence of an antibody bound to, for example, CDR3. However, this antibody blocks the approach of further copies of CD4, preventing interaction with other gp120 molecules within the virus spike oligomer. Higher-order complexes between gp120 and CD4 are thereby prevented and so virus-cell fusion does not occur. We have found that antibodies to rat CD4 domain 1 CDR3 region, and some to domain 2 and domains 3/4 are inhibitory to infection of cells expressing rat/human chimeric CD4 by HIV even though none of these epitopes are shared with human CD4 (Simon et al. unpublished data). In addition, several epitopes in domains 2, 3 and 4 of rat CD4 were masked following interaction with oligomeric, but not monomeric, gp120. These results are consistent with the notion that there is a concerted interaction between two or more CD4 monomers and gp120 oligomers during the infection process.

## 2.5 gp120: A CD4 Ligand

The studies discussed above suggested that, in the virus, the envelope glycoproteins may exist as oligomeric structures and that the physiologically relevant interactions are between oligomeric gp120/gp41 and CD4, rather than between the free monomers. More direct evidence has come from sucrose gradient studies using a soluble form of the envelope glycoprotein and either sCD4 or native CD4 released by detergent from cells in which evidence of a predominant complex consisting of two molecules of sCD4 complex with two copies of gp120/gp41 is presented (MALVOISIN and WILD 1994). The three-dimensional structure of gp120 has not yet been determined. The primary sequence consists of alternating segments of relatively conserved residues (C1–C5) alternating with disulphide-delimited loops (V1–V5) that show substantial sequence variation between strains (LEONARD et al. 1990). By exploiting a growing panel of mAbs against mapped continuous epitopes of gp120, it appears that extensive portions of the molecule that are exposed in the monomeric form, including regions C1 and C5, are occluded in the oligomer (MOORE et al. 1994).

### 2.5.1 The CD4 Binding Site

Residues of gp120 involved in CD4 binding have been identified using a combination of mAbs and site-directed mutagenesis (NYGREN et al. 1988; CORDONNIER et al. 1989). These have indicated that the binding site is discontinuous, with residues in C2, V4, C4 and C5, combining to form a contact site (OLSHEVSKY et al. 1990; SYU et al. 1990; THALI et al. 1991). Several regions of gp120, including the $NH_2$-terminal 62 amino acids, COOH-terminal 20 amino acids, the V1, V2 and V3 regions can be removed without significantly affecting CD4 affinity (POLLARD et al. 1992).

The CD4 binding site is not invariant. Much higher concentrations of sCD4 were required to neutralize the infectivity of HIV-1 strains that were capable of growing in primary macrophages, compared to those that had been adapted to growth in T cell lines in vitro (GOMATOS et al. 1990). Laboratory variants with lower affinities for CD4 were also more resistant to sCD4 neutralization (MCKEATING et al. 1991b; THALI et al. 1991) and, although their monomeric gp120 has a similar affinity for CD4 to that of the wild type, their oligomeric, virion-bound gp120 has a substantially lower affinity (MOORE et al. 1992a).

### 2.5.2 CD4 Induced Dissociation of gp120 from gp41

A second mechanism of sCD4-mediated neutralization is the induction of gp120-gp41 dissociation at phsysiological temperatures (MOORE et al. 1990, 1991). Strains of HIV-1 that varied in their sensitivity to sCD4 neutralization were correspondingly sensitive to sCD4-mediated gp120 dissociation (MCKEATING et al. 1991a; ORLOFF et al. 1993), determined by amino acid changes in the portion of gp41 that is believed to contact gp120 (MOORE et al. 1993) or the CD4-binding region of gp120 and the fusogenic peptide of gp41 (WILLEY et al. 1994). The physiological relevance of this process is open to question. It has been shown that the physicochemical characteristics of dissociation and of syncytium formation are quite different (FU et al. 1993), although the relationship between syncytium formation and virus-cell fusion is not straightforward. Dissociation may be an extreme manifestation of a multistep process involving conformational changes in the envelope glycoprotein following CD4 binding (SATTENTAU et al. 1993) which prepare the transmembrane protein for interaction with the plasma membrane of the host cell. Changes in the envelope glycoproteins that enhance this process seem to be selected by in vitro passage in T cell lines, because they tend to shorten the replicative cycle by allowing rapid entry.

### 2.5.3 Role of Variable Loops in Infection

The variable loops of gp120 are not only associated with tropism-determining interstrain variation, they are also the major neutralization epitopes. Antibodies that bind V3, for example, do not block attachment to the cell but prevent subsequent steps in viral entry (reviewed by CHENG-MAYER 1990). Mutants of gp120 in which these loops have been deleted still bind CD4 and their CD4 binding site is more exposed, but the resulting virus is noninfectious (WYATT et al. 1993). There is a possibility that V3 is the ligand for a cellular co-receptor (see Sect. 3) and the two receptor-ligand interactions may cooperate during viral entry (MCKEATING et al. 1992; POTTS et al. 1993), although it has been demonstrated that observed synergy between sCD4 and anti-V3 loop antibodies may be an artefact of the conditions used (SPOUGE 1994). In conclusion, the variable loops have an active, though as yet undefined, role during the post-attachment entry process.

# 3 Non-CD4 Components of the HIV Receptor

Although the function of CD4 as the principal component of cellular permissivity to HIV infection seems to be clear, there is compelling evidence that CD4 is neither sufficient nor always necessary for viral entry.

## 3.1 Evidence for Coreceptors and Alternative Receptors

### 3.1.1 A Block to Infection in nonprimate Cells

Initial studies showed that expression of CD4 in murine cells did not render them susceptible to HIV infection and that the block was after virion attachment but before envelope-plasma membrane fusion (MADDON et al. 1986). This suggested that there might be a molecule in addition to CD4 that was required for penetration into the cytoplasm and which was not expressed by rodent cells. Although the binding site for HIV-1 on CD4 is also utilized by HIV-2 and SIV (SATTENTAU et al. 1988), these three viruses show differential ability to replicate in a number of primate cell lines transfected with human CD4, suggesting that the non-CD4 coreceptor might be different in each case (CLAPHAM et al. 1991).

### 3.1.2 CD4-Independent Infection

A number of human cell lines which are apparently CD4-negative show a low sensitivity to HIV-1 infection (reviewed by CLAPHAM 1991). For example, HIV can infect CD4-negative human cells such as the glial cells of the brain (CHRISTOFINIS et al. 1987) and a variety of other human primary cells and lines (CHESEBRO et al. 1990; MIZRACHI et al. 1991), albeit at very low levels. Transfection of some of these lines with CD4 does not increase their permissivity for HIV (CHESEBRO et al. 1990; BEDDOWS et al. 1993). Some, such as Langerhans cells, are CD4-positive but remain susceptible when CD4 is removed (SCHMITT et al. 1990); others, such as U373-MG, remain insusceptible to HIV-1 even after CD4 transfection (CHESEBRO et al. 1990; HARRINGTON and GEBALLE 1993). These results suggested that there were alternative receptors to CD4 that could sometimes act independently of CD4. This sort of study must be treated with caution, however. With several of these lines infection can be blocked by pretreating the cells with anti-CD4 antibodies. It is therefore important to determine whether low levels of CD4 expression may account for the permissiveness of non hematopoietic cells.

In the case of HIV-2, requirements for CD4 seem to be less stringent. Some human cell lines are readily infectable and sensitive to cell fusion by a variant of HIV-$2_{ROD}$ independent of CD4 (CLAPHAM et al. 1992). Infection and fusion are not blocked by anti-CD4 or by soluble CD4. Indeed, soluble CD4 enhances the infection of these cells by other strains of HIV-2. It appears likely that soluble CD4 triggers conformational changes in the HIV-2 envelope spikes which allow better interaction with the non-CD4 receptor. Soluble CD4-induced enhancement of infection by SIV and HIV-2 of some CD4+ T cell lines has also been observed (ALLAN et al. 1990). This phenomenon also appears to be via an alternative receptor, as pretreatment of the cells by anti-CD4 does not block infection by the HIV-$2_{ROD}$ variant or by HIV treated with sCD4 (P.R.Clapham, and A. McKnight, personal communication).

### 3.1.3 Cell Type Tropism

The myriad strains of HIV that have been cultured in vitro have a wide range of tropisms for cell type: from those that grow preferentially in primary macrophages to those that grow rapidly in some or all immortalized T lymphoma cell lines (see WEISS 1933 and references cited by COLLIN et al. 1994 for a review of the field). The restriction to infection in nonpermissive cases is at the level of virus-cell fusion, rather than attachment, and is determined by sequences within the viral *env* gene, predominantly but not exclusively within the hypervariable V3 loop (e.g.HWANG et al. 1991)). It is believed that a number of cellular accessory factors that are differentially expressed on different cell types may be the receptors for the strain-specific envelope determinants (STEFANO et al. 1993).

It was possible that a single molecule or family of molecules would prove to serve as coreceptor, alternative receptor and tropism determinant, providing a unified solution to these three areas of investigation. However, studies to date seem to show that the picture is more complex.

## 3.2 Exceptions to Restriction in Nonprimate Cells

Not only may strains of HIV-1 with different cellular tropism use variant accessory factors but related primate lentiviruses may use a different, if overlapping, range of accessory factors. For example, the U87/CD4 human cell line is permissive to HIV-2 but not HIV-1; some rabbit and cat cell lines expressing human CD4 can be fused by HIV-2 and SIV but not HIV-1; and primate cells expressing human CD4 can be induced to fuse by SIV but not HIV-1 or HIV-2 (CLAPHAM et al. 1991; DRAGIC and ALIZON 1993; MCKNIGHT et al. 1994). Interestingly, human CD4-expressing rodent cells can be rendered HIV-2-susceptible by treatment with the membrane-altering agent amphotericin B (MCKNIGHT et al. 1994), suggesting that either the requirement for the accessory molecule is not absolute or that it is present but usually masked in these cells. It is now clear that even the accessory factor for HIV-1 is encoded by rodents, as one rat cell line, the myeloma Y3, is fully susceptible to HIV-1 entry after transfection with human CD4 (SIMON et al. 1994). This shows that the accessory factor cannot be exclusively the primate variant of one or more cell surface glycoproteins, as has been assumed.

Some cell types are sensitive to cell-to-cell fusion mediated through CD4-gp120 interaction while remaining resistant to fusion and penetration by cell-fee virus. This has been observed for HIV-2 on rabbit cells expressing human CD4 (MCKNIGHT et al. 1994) and for some T cell line tropic strains of HIV-1 (e.g.RF) on human monocyte-derived macrophages (G. Symmons, A. McKnight and P.R. Clapham, personal communication).

## 3.3 Studies on Heterokaryons and Genetic Complementation

One way of getting a handle on the non-CD4 component of the HIV receptor is to use genetic complementation by whole genomes, chromosomes or individual genes to determine the complexity of the genetics of susceptibility. The block to HIV entry in murine cells expressing human CD4 is not the result of modulation of CD4 itself and, in interspecific cell hybridomas, it appears that fewer than five human chromosomes can complement the murine defeat (WEINER et al. 1991). Heterokaryons made between human CD4-expressing murine and human cell lines and between susceptible and insusceptible human cells all support HIV infection (DRAGIC et al. 1992; BRODER et al. 1993; DRAGIC and ALIZON 1993; HARRINGTON and GEBALLE (1993). Consequently, the blocks to infection in each restricted case are the result of a lack of expression of essential genes rather than expression of some HIV inhibitory gene. We have observed that Y3, the rodent cell line that naturally expresses the HIV-1 accessory factor, forms spontaneous heterokaryons with L929/hCD4 cells when mixed in vitro. These heterokaryons are susceptible to HIV entry (SIMON and JAMES 1994), showing that both the rodent and human genes encoding the accessory factor can complement insusceptible cells in *trans*.

Recently, it has been shown that the accessory factor is found in the plasma membranes of human, but not murine or ovine, erythrocytes. When erythrocyte ghosts are fused with murine cells expressing human CD4, the resultant cells can fuse with HIV-1-infected cells in a CD4-dependent manner (DRAGIC et al. 1995). This phenotypic complementation is resistant to proteinase K and pronase treatment of the RBC ghosts, suggesting that the accessory factor might not be a protein.

A number of attempts have been made to isolate the gene for the accessory factor from cDNA libraries of susceptible cells and, while susceptibility is found in cells transfected with unfractionated libraries, attempts to isolate individual complementing cDNAs have been unsuccessful to date. This has led many to speculate that either the accessory factor is synthesized by the products of multiple genes or that there are multiple, essential accessory factors.

## 3.4 Candidate Coreceptors and Accessory Molecules

### 3.4.1 Glycolipids

The sphingolipid galactosyl ceramide (gal-cer), characteristically abundant in the central nervous system, is expressed on a number of HIV-susceptible CD4-negative cell lines of the brain and gut. Antibodies to gal-cer inhibit HIV infection of these cell lines and gp120 binds to gal-cer, but not other glycolipids, on thin layer chromatograms (Harouse et al. 1991) and to sulphatide, a related molecule (van den Berg et al. 1992). It was claimed that the binding site for gal-cer and sulphatide lay within residues 206–275 of gp120 (Bhat et al. 1993), but this region has been shown to bind to liposomes nonspecifically and have no selectivity for gal-cer (Berson and Doms, personal communication). A more recent report shows that the binding site is in or near the V3 loop (Cook et al. 1994), and subsequently, the same authors have used virus envelope chimeras to confirm that the gal-cer tropism of HIV-1 strains is determined by the COOH-terminal half of gp120, which contains the V3 and CD4 binding regions (Harouse and Gonzalez-Scarano, personal communication). It has been claimed that gp120 binds to the ganglioside GM1 on TLC plates (McAlarney et al. 1994) and to gal-cer and sulphatide. The involvement of these glycolipids in HIV infection of CD4-positive cells is doubted, as antibodies to gal-cer, while reducing virion binding, do not inhibit infection in macrophages (Seddiki et al. 1994).

### 3.4.2 Sulphated Polysaccharides

Dextran sulphate and certain other sulphated polysaccharides inhibit HIV replication by blocking attachment of virions to cell surface CD4 (Ito et al. 1987; Baba et al. 1988a; Mitsuya et al. 1988). There are conflicting claims that they inhibit or do not inhibit the binding of soluble gp120 to CD4 (Lederman et al. 1989; Callahan et al. 1991). There is evidence that they bind to the primary receptor for HIV, CD4, at a site near the gp120 binding region (Lederman et al. 1989; Parish et al. 1990). However, they have been shown to bind gp120, particularly the V3 region, and this interaction is inhibited by CD4 CDR3-derived peptides (Schols et al. 1990; Batinic and Robey 1992; Harrop et al. 1994). The interaction is electrostatic (Meshcheryakova et al. 1993), specific, but low-affinity ($K_d$ around $10^{-4}$ *M*; Mbemba et al. 1992a, b). Sulphated polysaccharides show inhibitory effects against a variety of enveloped viruses and a degree of selectivity (Baba et al. 1988b; McClure et al. 1992; Tanabe et al. 1992) based on preference for certain sequences of amino acid residues in the envelope glycoproteins (Hosoya et al. 1991). In the case of Herpes virus type-1, the initial attachment of the virus to susceptible cells is via heparan sulphate (see review by SPEAR et al. 1992).

It has been proposed that HIV also uses cell surface proteoglycans as a component of its cellular receptor and that dextran sulphate-V3 loop interactions mimic an essential postbinding interaction of HIV with the host cell (CALLAHAN et al. 1991). In support of this, purified heparan sulphate also inhibits HIV infection and treatment of cells with heparinase and chlorate but not chondroitinase reduced multiround replication of HIV and virus binding (PATEL et al. 1993). However, we have found that depletion of cell surface sulphation of glycosaminoglycans has no effect upon the efficiency of entry of HIV into susceptible cells, in contrast to HSV-1 (unpublished data).

### 3.4.3 Cell Surface Glycoproteins

The leukocyte adhesion receptor LFA-1 has been proposed to be involved in HIV-mediated cell fusion and infection as a result of studies using anti-LFA-1 antibodies (HILDRETH and ORENTAS 1989; GOUDSMIT and SMIT 1990; VALENTIN et al. 1990; VERMOTDESROCHES et al. 1991) but was shown not to be the rate-limiting factor under usual infection conditions in which gp120 concentrations are locally high (GRUBER et al. 1991). As the effects of LFA-1 antibodies are greater against syncytium formation than against virus infection, they could be acting simply to reduce cellular clumping, thereby indirectly reducing the frequency of syncytia. Because of its intimate involvement with CD4, a role for MHC class II has been proposed (MANN et al. 1988), and evidence for the involvement of CD7 has been presented (SATO et al. 1994).

A more direct approach has been to identify cell surface proteins that interact with gp41, the transmembrane subunit of the envelope glycoprotein that carries an $NH_2$-terminal hexapeptide that is responsible for virus-cell fusion (OWENS et al. 1990). Using a combination of affinity chromatography and western blotting with gp41, four candidate cellular proteins were identified, of which three could partially inhibit gp41 binding to HIV-susceptible cells (CHEN et al. 1993). It is too early to tell whether these proteins comprise the accessory factor.

### 3.4.4 Proteases and Peptidases

The hypervariable V3 loop of gp120, which is the principal neutralizing epitope and helps to determine the cell type tropism of the virus, contains a relatively conserved, exposed β-turn of sequence GPGQ (40% of strains sequenced), GPGR (35%), GPGK (4%), GLGQ (4%), GWGR (3%) and 17 other minor variants. A membrane-bound serine proteinase, tryptase TL2, which binds gp120, cleaves the V3 loop and can be inhibited by GPGR-containing peptides, has been isolated from the HIV-susceptible line MOLT-4 (KIDO et al. 1991). A similar though distinct serine proteinase has been isolated from U937 cells (AVRIL et al. 1993). Although some inhibitors of serine proteinases inhibit HIV-mediated syncytium formation, no conclusive proof that V3 cleavage is required for HIV entry into cells has been presented. Indeed, mutations in V3 that alter its protease susceptibility do not abrogate infectivity (SCHULTZ et al. 1993).

The dipeptidyl peptidase CD26 has been reported to be sufficient to render human CD4-expressing murine cells permissive for HIV (CALLEBAUT et al. 1993), though others have noted the lack of CD26 substrate sites in gp120 and gp41, its tissue distribution, the failure of anti-CD26 antibodies to inhibit infection and the failure of CD26 transfection to render murine cells permissive (BRODER et al. 1994). Consequently, the role of CD26 in HIV infection is still open to doubt.

### 3.4.5 Fc and Complement Receptors

The ability of antibody and complement to enhance the infectivity of arboviruses for monocytes and macrophages is well attested and it was thought that this might also prove to be an alternative route for HIV infection. While it has been demonstrated that anti-envelope antibody and complement can mediate attachment of HIV virions to appropriate cells and produce very moderate levels of infectivity enhancement in vitro, the infection process can still be blocked by sCD4 and anti-CD4 antibodies (TAKEDA et al. 1988; JOUAULT et al. 1989; REISINGER et al. 1990; ZEIRA et al. 1990; JOUALT et al. 1991; JUNE et al. 1991; MONTEFIORI et al. 1992; THIEBLEMONT et al. 1993). Expression of cytomegalovirus-encoded complement receptors on cultured cells in the absence of CD4 can be lead to low-affinity infection (MCKEATING et al. 1990). Although it is clear that enhancement does not provide an alternative route for viral infection, it might be quantitatively important in vivo during the establishment of foci of infection in lymphoid tissues containing intimate mixtures of cells of the monocyte and lymphocyte lineages.

## 4 Conclusions

Different retroviruses utilize receptors belonging to many families of cell surface proteins (reviewed by WEISS 1993). For cells that become productively infected by HIV in vivo, the overwhelmingly important cellular receptor is CD4. A limited region of the CDR2-like portion of CD4 domain 1, centred on Phe-43 makes contact with the oligomeric viral envelope glycoprotein gp120. Mutations, peptides and antibodies that suggest the specific involvement of other regions usually have alternative explanations. For example, antibodies to other site may sterically hinder the multisubunit interactions between CD4 and oligomeric gp120 that are required for infection or, as in the case of the D2-D3 hinge region, implicate a conserved structural feature that needs to remain flexible but tolerates substantial sequence variation.

Following the multimolecular interaction between CD4 and gp120, conformational changes occur in the envelope glycoproteins allowing the V3 loop of gp120, the fusogenic peptide of gp41 or possibly other sites to make contact with accessory factors on the surface of the cell, leading to fusion between cellular and viral membranes. Circumstantial evidence suggests that the accessory factors are either directly encoded by several genes or are synthesized by at least two gene products and these genes are encoded in both the human and murine genomes. However, while most human cells express both or all of these genes simultaneously, most murine cells do not. It is possible that HIV-2, SIV and the strains of HIV with varying cellular tropism use related coreceptors that are the product of nonidentical, though possibly overlapping, sets of genes. These considerations should be borne is mind when considering the candidates that have been proposed to be coreceptors for HIV and suggest that it might be worth investigating small animals transgenic for human CD4 as models for the infection and pathogenesis of HIV-2.

Of the candidate molecules proposed to date, only glycolipids appear to fulfil most of these criteria. These molecules come in a large variety of glycoforms, and the glycosyl transferases and glycosidases that are responsible for their biosynthesis from and their degradation to ceramide are encoded in the genomes of most species. It is known that different members of the galactosyl, globo and lactoneo series are present in widely differing abundance on different cell types, in different species, and at different stages of activation, differentiation and malignant transformation and thus have the potential to act as the cellular determinants of viral tropism in the guise of accessory molecules for HIV infection. However, the techniques for analysing these molecules are much less sensitive and powerful than those for investigating nucleic acids and proteins, and so it might take some time to ascertain their importance. A low-affinity interaction between the envelope glycoproteins of a cell-bound HIV and an abundant cell surface molecule may be sufficient to facililate fusion and it is notoriously difficult to identify specificity in such interactions.

Although all the primate immunodeficiency viruses use CD4 as the principal receptor (Sattentau et al. 1988), the lentiviruses of ungulates as well as FIV do not. It is conceivable that the putative accessory factors were originally the principal, albeit low-affinity, lentivirus receptor and that the adoption of CD4 as a higher-affinity receptor may be a relatively recent event in lentivirus evolution.

## References

Allan J, Strauss J, Buck D (1990) Enhancement of SIV infection with soluble receptor molecules. Science 247: 1084–1088

Arthos J, Deen KC, Chaikin MA, Fornwald JA, Sathe G, Sattentau QJ, Clapham PR, Weiss RA, McDougal JS, Pietropaolo C, Axel R, Truneh A, et al. (1989) Identification of the residues in human CD4 critical for the binding of HIV. Cell 57: 469–481

Ashkenazi A, Presta LG, Marsters SA, Camerato TR, Rosenthal KA, Fendly BM, Capon DJ (1990) Mapping the CD4 binding site for human immunodeficiency virus by alanine-scanning mutagenesis. Proc Natl Acad Sci USA 87: 7150–7154

Autiero M, Abrescia P, Dettin M, Di BC, Guardiola J (1991) Binding to CD4 of synthetic peptides patterned on the principal neutralizing domain of the HIV-1 envelope protein. Virology 185: 820–828

Avril LE, Di MM, Barin F, Gauthier F (1993) Interaction between a membrane-associated serine proteinase of U–937 monocytes and peptides from the V3 loop of the human immunodeficiency virus type 1 (HIV-1) gp120 envelope glycoprotein. FEBS Lett 317: 167–172

Baba M, Pauwels R, Balzarini J, Arnout J, Desmyter J, De CE (1988a) Mechanism of inhibitory effect of dextran sulfate and heparin on replication of human immunodeficiency virus in vitro. Proc Natl Acad Sci USA 85: 6132–6136

Baba M, Snoeck R, Pauwels R, De CE (1988b) Sulfated polysaccharides are potent and selective inhibitors of various enveloped viruses including herpes simplex virus cytomegalovirus, vesicular stomatitis virus, and human immunodeficiency virus. Antimicrob Agents Chemother 32: 1742–1745

Batinic D, Robey FA (1992) The V3 region of the envelope glycoprotein of human immunodeficiency virus type 1 binds sulfated polysaccharides and CD4-derived synthetic peptides. J Biol Chem 267: 6664–6671

Beddows S, Bieniasz P, Shaunak S, Weber J (1993) HIV replication in CD4-negative cell lines: Effect of cloning, CD4 expression and inhibitory by dextrin sulphate. Antiviral Chem Chemother 4: 173–177

Berger EA, Lifson JD, Eiden LE (1991) Stimulation of glycoprotein gp120 dissociation from the envelope glycoprotein complex of human immunodeficiency virus type 1 by soluble CD4 and CD4 peptide derivatives: implications for the role of the complementarity-determining region 3-like region in membrane fusion. Proc Natl Acad Sci USA 88: 8082–8086

Bhat S, Mettus RV, Reddy EP, Ugen KE, Srikanthan V, Williams WV, Weiner DB (1993) The galactosyl ceramide/sulfatide receptor binding region of HIV-1 gp120 maps to amino acids 206–275. AIDS Res Hum Retroviruses 9: 175–181

Brady RL, Dodson EJ, Dodson GG, Lange G, Davis SJ, Williams AF, Barclay AN (1993) Crystal structure of domains 3 and 4 of rat CD4: relation to the $NH_2$-terminal domains. Science 260: 976–983

Broder CC, Berger EA (1993) CD4 molecules with a diversity of mutations encompassing the CDR3 region efficiently support human immunodeficiency virus type 1 envelope glycoprotein-mediated cell fusion. J Virol 67: 913–926

Broder CC, Dimitrov DS, Blumenthal R, Berger EA (1993) The block to HIV-1 envelope glycoprotein-mediated membrane fusion in animal cells expressing human CD4 be overcome by a human cell component(s). Virology 193: 483–491

Broder CC, Nussbaum O, Gutheil WG, Bachovchin WW, Berger EA, Patience C, McKnight A, Clapham PR, Boyd MT, Weiss RA, Schulz TF, Camerini D et al. (1994) CD26 antigen and HIV fusion? Science 264: 1156–1165

Brodsky MH, Warton M, Myers RM, Littman DR (1990) Analysis of the site in CD4 that binds to the HIV envelope glycoprotein. J Immunol 144: 3078–3086

Burkly LC, Olson D, Shapiro R, Winkler G, Rosa JJ, Thomas DW, Williams C, Chisholm P (1992) Inhibition of HIV infection by a novel CD4 domain 2-specific monoclonal antibody: dissecting the basis for its inhibitory effect on HIV-induced cell fusion. J Immunol 149: 1779–1787

Callahan LN, Phelan M, Mallinson M, Norcross MA (1991) Dextran sulfate blocks antibody binding to the principal neutralizing domain of human immunodeficiency virus type 1 without interfering with gp120–CD4 interactions. J Virol 65: 1543–1550

Callebaut C, Krust B, Jacotot E, Hovanessian AG (1993) T cell activation antigen, CD26, as a cofactor for entry of HIV in $CD4^+$ cells. Science 262: 2045–2050

Camerini D, Seed B (1990) A CD4 domain important for HIV-mediated syncytium formation lies outside the virus binding site. Cell 60: 747–754

Cereda PM, Palu G, Rassu M, Toni M, Malwood W, Dettin M, Di BC (1991) Anti-HIV-1 activity of CD4 synthetic oligopeptides representative of the putative gp120 binding site. Antiviral Chem Chemother 2: 157–161

Chen YH, Bock G, Vornhagen R, Steindl F, Katinger H, Dierich MP (1993) HIV-1 gp41 binds to several proteins on the human B-cell line, Raji. Mol Immunol 30: 1159–1163

Cheng-Mayer C (1990) Biological and molecular features of HIV-1 related to tissue tropism. AIDS 4: S49–S56

Chesebro B, Buller R, Portis J, Wehrly K (1990) Failure of human immunodeficiency virus entry and infection in CD4-positive human brain and skin cells. J Virol 64: 215–221

Choe HR, Sodroski J (1992) Contribution of charged amino acids in the CDR2 region of CD4 to HIV-1 gp120 binding. J Acquir Immune Defic Syndr 5: 204–210

Christofinis G, Papadaki L, Sattentau Q, Ferns RB, Tedder R (1987) HIV replicates in cultured human brain cells. AIDS 1: 229–234

Clapham P (1991) Human immunodeficiency virus infection of non-haematopoietic cells. The role of CD4-independent entry. Rev Med Virol 1: 51–58

Clapham P, McKnight A, Weiss R (1992) Human immunodeficiency virus Type2 infection and fusion of CD4-negative cell lines: induction and enhancement by soluble CD4. J Virol 66: 3531–3537

Clapham PR, Blanc D, Weiss RA (1991) Specific cell surface requirements for the infection of CD4-positive cells by human immunodeficiency virus types 1 and 2 and by Simian immunodeficiency virus. Virology 181: 703–15

Clayton LK, Hussey RE, Steinbrich R, Ramachandran H, Husain Y, Reinherz EL (1988) Substitution of murine for human CD4 residues identifies amino acids critical for HIV-gp120 binding. Nature 335: 363–366

Collin M, Illei P, James W, Gordon S (1994) Definition of the range and distribution of HIV macrophage tropism using PCR-based infectivity measurements J Gen Virol 75: 1597–1603

Cook DG, Fantini J, Spitalnik SL, Gonzalez-Scarano F (1994) Binding of human immunodeficiency virus type I (HIV-1) Gp120 to galactosylceramide (GalCer): relationship to the V3 Loop. Virology 201: 206–214

Cordonnier A, Montagnier L, Emerman M (1989) Single amino-acid changes in HIV envelope affect viral tropism and receptor binding. Nature 340: 571–574

Dalgleish AG, Beverley PC, Clapham PR, Crawford DH, Greaves MF, Weiss RA (1984) The CD4 (T4) antigen is an essential component of the receptor for the AIDS retrovirus. Nature 312: 763–767

Dalgleish AG, Chanh TC, Thomson BJ, Malkovsky M, Kennedy RC (1987) Neutralisation of HIV isolates by anti-idiotypic antibodies which mimic the T4 (CD4) epitope: a potential AIDS vaccine. Lancet 2: 1047–1049

Davis SJ, Schockmel GA, Somoza C, Buck DW, Healey DG, Rieber EP, Reiter C, Williams AF (1992) Antibody and HIV-1 gp120 recognition of CD4 undermines the concept of mimicry between antibodies and receptors. Nature 358: 76–79

Dragic T, Alizon M (1993) Different requirements for membrane fusion mediated by the envelopes of human immunodeficiency virus types 1 and 2. J Virol 67: 2355–2359

Dragic T, Charneau P, Clavel F, Alizon M (1992) Complementation of murine cells for human immunodeficiency virus envelope/CD4-mediated fusion in human/murine heterokaryons. J Virol 66: 4794–4802

Dragic T, Picard L, Alizon M (1995) Proteinase-resistant factors in human erythrocyte membranes mediate CD4-dependent fusion with cells expressing HIV-1 envelope glycoproteins. J Virol 69: 1013–1018

Eiden LE, Lifson JD (1992) HIV interactions with CD4 : a continuum of conformations and consequences. Immunol Today 13: 201–206

Fisher RA, Bertonis JM, Meier W, Johnson VA, Costopoulos DS, Liu T, Tizard R, Walker BD, Hirsch MS, Schooley RT, Flavell RA (1988) HIV infection is blocked in vitro by recombinant soluble CD4. Nature 331: 76–78

Fu YK, Hart TK, Jonak ZL, Bugelski PJ, (1993) Physicochemical dissociation of CD4-mediated syncytium formation and shedding of human immunodeficiency virus type 1 gp120. J Virol 67: 3818–3825

Golding H, Blumenthal R, Manischewitz J, Littman DR, Dimitrov DS (1993) Cell fusion mediated by interaction of a hybrid CD4.CD8 molecule with the human immunodeficiency virus type 1 envelope glycoprotein does occur after a long lag time. J Virol 67: 6469–6475

Gomatos PJ, Stomatos NM, Gendelman HE, Fowler A, Hoover DL, Kalter DC, Burke DS, Tramont EC, Meltzer MS (1990) Relative inefficiency of soluble recombinant CD4 for inhibition of infection by monocyte-tropic HIV in monocytes and T cells. J Immunol 144: 4183–4188

Goudsmit J, Smit L (1990) CD 11a/CD18 (LFA-1) epitopes involved in syncytium formation among CD4+T-cells following cell free HIV-1 infection. Viral Immunol 3: 289–293

Gruber MF, Webb D, Gerrard TL, Mostowski HS, Vujcic L, Golding H (1991) Re-evaluation of the involvement of the adhesion molecules ICAM-1/LFA-1 in syncytia formation of HIV-1-infected subclones of a CEM T-cell leukemic line. AIDS Res Hum Retroviruses 7: 45–53

Harouse JM, Bhat S, Spitalnik SL, Laughlin M, Stefano K, Silbergerg DH, GonzalezScarano F (1991) Inhibition of entry of HIV-1 in neural cell lines by antibodies against galactosyl ceramide. Science 253: 320–323

Harrington RD, Geballe AP (1993) Cofactor requirement for human immunodeficiency virus type 1 entry into a CD4-expressing human cell line. J Virol 67: 5939–5947

Harrop HA, Coombe DR, Rider CC (1994) Heparin specifically inhibits binding of V3 loop antibodies to HIV-1 gp120, an effect potentiated by CD4 binding. AIDS 8: 183–192
Hasunuma T, Tsubota H, Watanabe M, Chen ZW, Lord CI, Burkly LC, Daley JF, Letvin NL (1992) Regions of the CD4 molecule not involved in virus binding or syncytia formation are required for HIV-1 infection of lymphocytes. J Immunol 148: 1841–1846
Healey D, Dianda L, Moore JP, McDougal JS, Moore MJ, Estess P, Buck D, Kwong PD, Beverley P, Sattentau QJ (1990) Novel anti-CD4 monoclonal antibodies separate human immunodeficiency virus infection and fusion of CD4+ cells from virus binding. J Exp Med 172: 1233–1242
Healey DG, Dianda L, Buck D, Schroeder K, Truneh A, Sattentau QJ, Beverley P (1991) A highly selected panel of anti-CD4 antibodies fails to induce anti-idiotypic antisera mediating human immunodeficiency virus neutralization. Eur J Immunol 21: 1491–1498
Hildreth J, Orentas RJ (1989) Involvement of a leukocyte adhesion receptor (LFA-I) in HIV-induced syncytium formation. Science 244: 1075–1078
Hosoya M, Balzarini J, Shigeta S, De CE (1991) Differential inhibitory effects of sulfated polysaccharides and polymers on the replication of various myxoviruses and retroviruses, depending on the composition of the target amino acid sequences of the viral envelope glycoproteins. Antimicrob Agents Chemother 35: 2515–2520
Hussey RE, Richardson NE, Kowalski M, Brown NR, Chang HC, Siliciano RF, Dorfman T, Walker B, Sodroski J, Reinherz EL (1988) A soluble CD4 protein selectively inhibits HIV replication and syncytium formation. Nature 331: 78–81
Hwang SS, Boyle TJ, Lyerly HK, Cullen BR (1991) Identification of the envelope V3 loop as the primary determinant of cell tropism in HIV-1. Science 253: 71–73
Ito M, Baba M, Sato A, Pauwels R, De CE, Shigeta S (1987) Inhibitory effect of dextran sulfate and heparin on the replication of human immunodeficiency virus (HIV) in vitro. Antiviral Res 7: 361–367
Jameson BA , Rao PE, Kong LI, Hahn BH, Shaw GM, Hood LE, Kent S (1988) Location and chemical synthesis of a binding site for HIV-1 on the CD4 protein. Science 240: 1335–1339
Joualt T, Chapuis F, Bahraoui E, Gluckman JC (1991) Infection of monocytic cells by HIV 1: Combined role of FcR and CD4. Res Virol 142: 183–188
Jouault T, Chapuis F, Olivier R, Parravicini C, Bahraoui E, Gluckman JC (1989) HIV infection of monocytic cells: role of antibody-mediated virus binding to Fc-gamma receptors. AIDS 3: 125–133
June RA, Schade SZ, Bankowski MJ, Kuhns M, McNamara A, Lint TF, Landay AL, Spear GT (1991) Complement and antibody mediate enhancement of HIV infection by increasing virus binding and provirus formation. AIDS 5: 269–274
Kalyanaraman VS, Rausch DM, Osborne J, Padgett M, Hwang KM, Lifson JD, Eiden LE (1990) Evidence by peptide mapping that the region CD4(81–92) is involved in gp120/CD4 interaction leading to HIV infection and HIV-induced syncytium formation. J Immunol 145: 4072–4078
Kido H, Fukutomi A, Katunuma N (1991) A novel membrane-bound serine esterase in human T4+-lymphocytes is a binding protein of envelope glycoprotein gp120 of HIV-1. Biomed Biochim Acta 50: 781–789
Klatzmann D, Barre SF, Nugeyre MT, Danquet C, Vilmer E, Griscelli C, Brun VF, Rouzioux C, Gluckman JC, Chermann JC et al. (1984a) Selective tropism of lymphadenopathy associated virus (LAV) for helper-inducer T lymphocytes. Science 225: 59–63
Klatzmann D, Champagne E, Chamaret S, Gruest J, Guetard D, Hercend T, Gluckman JC, Montagnier L (1984b) T-lymphocyte T4 molecule behaves as the receptor for human retrovirus LAV. Nature 312: 767–768
Langedijk J, Puijk WC, Van HW, Meloen RH (1993) Location of CD4 dimerization site explains critical role of CDR3-like region in HIV-1 infection and T-cell activation and implies a model for complex of coreceptor-MHC. J Biol Chem 268: 16875–16878
Lasarte JJ, Sarobe P, Golvano J, Prieto I, Civeira MP, Gullon A, Sarin PS, Prieto J, BorrasCuesta F (1994) CD4-modified synthetic peptides containing phenylalanine inhibit HIV-1 infection in vitro. J Acquir Immune Defic Syndr 7: 129–134
Layne SP, Merges MJ, Dembo M, Spouge JL, Nara PL (1990) HIV requires multiple gp120 molecules for CD4-mediated infection. Nature 346: 277–279
Lederman S, Gulick R, Chess L (1989) Dextran sulfate and heparin interact with CD4 molecules to inhibit the binding of coat protein (gp120) of HIV. J Immunol 143: 1149–1154
Leonard C, Spellman N, Riddles L, Harris R, Thomas J, Gregory T (1990) Assignment of intrachain disulphide bonds and characterization of potential glycosylation sites of type 1 human immunodeficiency virus envelope glycoprotein (gp120) expressed in chinese hamster ovary cells. J Biol Chem 265: 10373–10376

Lifson JD, Hwang KM, Nara PL, Fraser B, Padgett M, Dunlop NM, Eiden LE (1988) Synthetic CD4 peptide derivatives that inhibit HIV infection and cytopathicity. Science 241: 712–716

Lifson JD, Rausch DM, Kalyanaraman VS, Hwang KM, Eiden LE (1991) Synthetic peptides allow discrimination of structural features of CD4 (81–92) important for HIV-1 infection versus HIV-1-induced syncytium formation. AIDS Res Human Retroviruses 7: 521–527

Maddon PJ, Dalgleish AG, McDougal JS, Clapham PR, Weiss RA, Axel R (1986) The T4 gene encodes the AIDS virus receptor and is expressed in the immune system and the brain. Cell 47: 333–348

Malvoisin E, Wild F (1994) Analysis of the human immunodeficiency virus type 1 envelope protein interaction with the CD4 host cell receptor. J Gen Virol 75: 839–847

Mann DL, Read Connole E, Arthur LO, Robey WG, Wernet P, Schneider EM, Blattner WA, Popovic M (1988) HLA-DR is involved in the HIV-1v binding site on cells expressing MHC class II antigens. J Immunol 141: 1131–1136

Mbemba E, Chams V, Gluckman JC, Klatzmann D, Gattegno L (1992a) Molecular interaction between HIV-1 major envelope glycoprotein and dextran sulfate. Biochim Biophys Acta 1138: 62–67

Mbemba E, Czyrski JA, Gattegno L (1992b) The interaction of a glycosaminoglycan, heparin, with HIV-1 major envelope glycoprotein. Biochim Biophys Acta 1180: 123–129

McAlarney T, Apostolski S, Lederman S, Latov N (1994) Characteristics of HIV-1 gp120 glycoprotein binding to glycolipids. J Neurosci Res 37: 453–460

McClure MO, Moore JP, Blanc DF, Scotting P, Cook G, Keynes RJ, Weber JN, Davies D, Weiss RA (1992) Investigations into the mechanism by which sulfated polysaccharides inhibit HIV infection in vitro. AIDS Res Hum Retroviruses 8: 19–26

McDougal JS, Nicholson J, Cross GD et al. (1986) Binding of the human retrovirus HTLV-III/LAV/ARV/HIV to the CD4 (T4) molecule: conformation dependence epitope mapping, antibody inhibition, and potential for idiotypic mimicry. J Immunol 137: 2937–2944

McKeating JA, Griffiths PD, Weiss RA (1990) HIV susceptibility conferred to human fibroblasts by cytomegalovirus-induced Fc receptor. Nature 343: 659–661

McKeating J, Balfe P, Clapham P, Weiss RA (1991a) Recombinant CD4-selected human immunodeficiency virus type 1 variants with reduced gp120 affinity for CD4 and increased cell fusion capacity. J Virol 65: 477–4785

McKeating JA, McKnight A, Moore JP (1991b) Differential loss of envelope glycoprotein gp120 from virions of human immunodeficiency virus type 1 isolates: effects on infectivity and neutralization. J Virol 65: 852–860

McKeating JA, Cordell J, Dean CJ, Balfe P (1992) Synergistic interaction between ligands binding to the CD4 binding site and V3 domain of human immunodeficiency virus type 1 gp120. Virology 191: 732–742

McKnight A, Clapham PR, Weiss RA (1994) HIV-2 and SIV infection of nonprimate cell lines expressing human CD4: restrictions to replication at distinct stages. Virology 201: 8–18

Meshcheryakova D, Andreev S, Tarasova S, Sidorova M, Vafina M, Kornilaeva G, Karamov E, Khaitov R (1993) CD4-derived peptide and sulfated polysaccharides have similar mechanisms of anti-HIV activity based on electrostatic interactions with positively charged gp120 fragments. Mol Immunol 30: 993–1001

Mitsuya H, Looney DJ, Kuno S, Ueno R, WongStaal F, Broder S (1988) Dextran sulfate suppression of viruses in the HIV family: inhibition of virion binding to CD4+ cells.Science 240: 646–649

Mizrachi Y, Zeira M, Shahabuddin M, Li G, Sinangil F, Volsky DJ (1991) Efficient binding, fusion and entry of HIV1 into CD4-negative neural cells: a mechanism for neuropathogenesis in AIDS. Bull Inst Pasteur 89: 81–96

Mizukami T, Fuerst TR, Berger EA, Moss B (1988) Binding region for human immunodeficiency virus (HIV) and epitopes for HIV-blocking monoclonal antibodies of the CD4 molecule defined by site-directed mutagenesis. Proc Natl Acad Sci USA 85: 9273–9277

Moebius U, Clayton LK, Abraham S, Harrison SC, Reinherz EL (1992) The human immunodeficiency virus gp120 binding site on CD4: delineation by quantitative equilibrium and kinetic binding studies of mutants in conjunction with a high-resolution CD4 atomic structure. J Exp Med 176: 507–517

Montefiori DC, Zhou J, Shaff DI (1992) CD4-independent binding of HIV-1 to the B lymphocyte receptor CR2 (CD21) in the presence of complement and antibody. Clin Exp Immunol 90: 383–389

Moore JP (1993) A monoclonal antibody to the CDR-3 region of CD4 inhibits soluble CD4 inhibits soluble CD4 binding to virions of human immunodeficiency virus type 1. J Virol 67: 3656–3659

Moore JP, McKeating JA, Weiss RA, Sattentau QJ (1990) Dissociation of gp120 from HIV-1 virions induced by soluble CD4. Science 250: 1139–1142

Moore JP, McKeating JA, Norton WA, Sattentau QJ (1991) Direct measurement of soluble CD4 binding to human immunodeficiency virus type 1 virions: gp120 dissociation and its implications for virus-cell binding and fusion reactions and their neutralization by soluble CD4. Journal of Virology 65: 1133–1140

Moore JP, McKeating JA, Huang YX, Ashkenazi A, Ho DD (1992a) Virions of primary human immunodeficiency virus type 1 isolates resistant to soluble CD4 (sCD4) neutralization differ in sCD4 binding and glycoprotein gp120 retention from sCD4-sensitive isolates. J Virol 66: 235–243

Moore JP, Sattentau QJ, Klasse PJ, Burkly LC (1992b) A monoclonal antibody to CD4 domain 2 blocks soluble CD4-induced conformational changes in the envelope glycoproteins of human immunodeficiency virus type 1 (HIV-1) and HIV-1 infection of CD4+ cells. J Virol 66: 4784–4793

Moore JP, Burkly LC, Connor RI, Cao Y, Tizard R, Ho DD, Fisher RA (1993) Adaptation of two primary human immunodeficiency virus type 1 isolates to growth in transformed T cell lines correlates with alterations in the responses of their envelope glycoproteins to soluble CD4. AIDS Res Hum Retroviruses 9: 529–539

Moore JP, Sattentau QJ, Wyatt R, Sodroski J (1994) Probing the structure of the human immunodeficiency virus surface glycoprotein gp120 with a panel of monoclonal antibodies. J Virol 68: 469–484

Nara PL, Hwang KM, Rausch DM, Lifson JD, Eiden LE (1989) CD4 antigen-based antireceptor peptides inhibit infectivity of human immunodeficiency virus in vitro at multiple stages of the viral life cycle. Proc Natl Acad Sci USA 86: 7139–7143

Nygren A, Bergman T, Matthews T, Jornvall H, Wigzell H (1988) 95- and 25-kDa fragments of the human immunodeficiency virus envelope glycoprotein bind to the CD4 receptor. Proc Natl Acad Sci USA 85: 6543–6546

Ohki K, Kimura T, Jones IM, Morita F, Ikuta K (1994) Multiple effects of CD4 CDR3-related peptide derivatives showing anti-HIV-1 activity on HIV-1 gp120 functions. Vaccine 12: 343–350

Olshevsky U, Helseth E, Furman C, Li J, Haseltine W, Sodroski J (1990) Identification of individual human immunodeficiency virus type 1 gp120 amino acids important for CD4 binding. J Virol 64: 5701–5707

Orloff SL, Kennedy MS, Belperron AA, Maddon PJ, McDougal JS (1993) Two mechanisms of soluble CD4 (sCD4)-mediated inhibition of human immunodeficiency virus type 1 (HIV-1) infectivity and their relation to primary HIV-1 isolates with reduced sensitivity to sCD4. J Virol 67: 1461–1471

Owens RJ, Tanner CC, Mulligan MJ, Srinivas RV, Compans RW (1990) Oligopeptide inhibitors of HIV-induced syncytium formation. AIDS Res Hum Retroviruses 6: 1289–1296

Pal R, Nair BC, Hoke GM, Sarngadharan MG, Edidin M (1991) Lateral diffusion of CD4 on the surface of a human neoplastic T-cell line probed with a fluorescent derivative of the envelope glycoprotein (gp120) of human immunodeficiency virus type 1 (HIV-1). J cell Physiol 147: 326–332

Parish CR, Low L, Warren HS, Cunningham AL (1990) A polyanion binding site on the CD4 molecule. Proximity to the HIV-gp120 binding region. J Immunol 145: 1188–1195

Patel M, Yanagishita M, Roderiquez G, Bou HD, Oravecz T, Hascall VC, Norcross MA (1993) Cell-surface heparan sulfate proteoglycan mediates HIV-1 infection of T-cell lines. AIDS Res Hum Retroviruses 9: 167–174

Pollard S, Rosa M, Rosa J, Wiley D (1992) Truncated variants of gp120 bind CD4 with high affinity and suggest a minimum CD4 binding region. EMBO J 11: 585–591

Potts BJ, Field KG, Wu Y, Posner M, Cavacini L, WhiteScharf M (1993) Synergistic inhibition of HIV-1 by CD4 binding domain reagents and V3-directed monoclonal antibodies. Virology 197: 415–419

Poulin L, Evans LA, Tang SB, Barboza A, Legg H, Littman DR, Levy JA (1991) Several CD4 domains can play a role in human immunodeficiency virus infection in cells. J Virol 65: 4893–4901

Rausch DM, Hwang KM, Padgett M, Voltz AH, Rivas A, Engleman E, Gaston I, McGrath M, Fraser B, Kalyanaraman VS, Nara PL, Dunlop N et al. (1990) Peptides derived from the CDR3-homologous domain of the CD4 molecule are specific inhibitors of HIV-1 and SIV infection, virusinduced cell fusion, and postinfection viral transmission in virtro: implications for the design of small-peptide anti-HIV therapeutic agents. Ann N Y Acad Sci 616: 125–148

Rausch DM, Lifson JD, Padgett MP, Chandrasekhar B, Lendvay J, Hwang KM, Eiden LE (1992) CD4(81–92)-based peptide derivatives; Structural requirements for blockade of HIV infection, blockade of HIV-induced syncytium formation, and virostatic activity in vitro. Biochem Pharmacol 43: 1785–1796

Reisinger EC, Vogetseder W, Berzow D, Kofler D, Bitterlich G, Lehr HA, Wachter H, Dierich MP (1990) Complement-mediated enhancement of HIV-1 infection of the monoblastoid cell line U937. AIDS 4: 961–965

Repke H, Gabuzda D, Palu G, Emmrich F, Sodroski J (1992) Effects CD4 synthetic peptides on HIV type I envelope glycoprotein function. J Immunol 149: 1809–1816

Richardson N, Brown N, Hussey R, Vaid A, Matthews T, Bolognesi D, Reinherz E (1988) Binding site for human immunodeficiency virus coat protien gp120 is located in the $NH_2$-terminal region of T4 (CD4) and requires the intact variable-region-like domain. Proc Natl Acad Sci USA 85: 6102–6106

Ryu SE, Kwong PD, Truneh A, Porter TG, Arthos J, Rosenberg M, Dai X, Xuong N, Axel R, Sweet RW, Hendrickson WA (1990) Crystal structure of an HIV-binding recombinant fragment of human CD4. Nature 348: 419–426

Sato AI, Balamuth FB, Ugen KE, Williams WV, Weiner DB (1994) Identification of CD7 glycoprotein as an accessory molecule in HIV-1- mediated syncytium formation and cellfree infection. J Immunol 152: 5142–5152

Sattentau QJ, Dalgleish AG, Weiss RA, Beverley P (1986) Epitopes of the CD4 antigen and HIV infection. Science 234: 1120–1123

Sattentau QJ, Clapham PR, Weiss RA, Beverley P, Montagnier L, Alhalabi MF, Gluckmann JC, Klatzmann D (1988) The human and simian immunodeficiency viruses HIV-1, HIV-2, and SIV interact with similar epitopes on their cellular receptor, the CD4 molecule. AIDS 2: 101–105

Sattentau QJ, Moore JP, Vignaux F, Traincard F, Poignard P (1993) Conformational changes induced in the envelope glycoproteins of the human and simian immunodeficiency viruses by soluble receptor binding. J Virol 67: 7383–7393

Schmitt D, DezutterDambuyant C, Hanau D, Kolbe H, Kieny MP, Cazenave JP, Thivolet J (1990) In vitro binding and internalization of HIV envelope glycoproteins by human epidermal Langerhans cells does not require the CD4-GP120-binding site. Res Virol 141: 209–215

Schockmel G, Somoza C, Davis S, Williams A, Healey D (1992) Construction of a binding site for human immunodeficiency virus type 1 gp120 in rat CD4. J Exp Med 175: 301–304

Schols D, Pauwels R, Desmyter J, de CE (1990) Dextran sulfate and other polyanionic anti-HIV compounds specifically interact with the viral gp120 glycoprotein expressed by T-Cells persistently infected with HIV-1. Virology 175: 556–561

Schultz T, Reeves J, Hoad J, Tailor C, Stephens P, Clements G, Ortlepp S, Page K, Moore J, Weiss R (1993) Effects of mutations in the V3 loop of HIV-1 gp120 on infectivity and susceptibility to proteolytic cleavage. AIDS Res Hum Retroviruses 9: 159–166

Seddiki N, Ramdani A, Saffar L, Portoukalian J, Gluckman JC, Gattegno L (1994) A monoclonal antibody directed to sulfatide inhibits the binding of human immunodeficiency virus type 1 (HIV-1) envelope glycoprotein to macrophages but not their infection by the virus. Biochim Biophys Acta Mol Basis Dis 1225: 289–296

ShapiraNahor O, Golding H, Vujcic LK, RestoRuiz S, Fields RL, Robey FA (1990) CD4-derived synthetic peptide blocks the binding of HIV-1 GP120 to CD4-bearing cells and prevents HIV-1 infection. Cell Immunol 128: 101–117

Simon J, Schockmel G, Illei P, James W (1994) A rodent cell line permissive for entry and reverse transcription of HIV-1 has a pre-integration block to productive infection. J Gen Virol 75: 2615–2623

Simon JH, Somoza C, Schockmel GA, Collin M, Davis SJ, Williams AF, James W (1993) A rat CD4 mutant containing the gp120-binding site mediates human immunodeficiency virus type 1 infection. J Exp Med 177: 949–954

Simon JHM, James W (1994) Heterokaryons formed between a rat myeloma and a mouse fibroblast are permissive for entry of HIV-1. AIDS Res Hum Retroviruses 70: 7609–7677

Smith DH, Byrn RA, Marsters SA, Gregory T, Groopman JE, Capon DJ (1987) Blocking of HIV-1 infectivity by a soluble, secreted form of the CD4 antigen. Science 238: 1704–1707

Spear PG, Shieh MT, Herold BC, WuDunn D, Koshy TI (1992) Heparan sulfate glycosaminoglycans as primary cell surface receptors for herpes simplex virus. Adv Exp Med Biol 313: 341–353

Spouge JL (1994) Viral multiplicity of attachment and its implications for human immunodeficiency virus therapies. J Virol 68: 1782–1789

Stefano KA, Collman R, Kolson D, Hoxie J, Nathanson N, GonzalezScarano F (1993) Replication of a macrophage-tropic strain of human immunodeficiency virus type 1 (HIV-1) in a hybrid cell line, CEMx174, suggests that cellular accessory molecules are required for HIV-1 entry. J Virol 67: 6707–6715

Sutor GC, Dreikhausen U, Vahning U, Jurkiewicz E, Hunsmann, G, Lundin K, Schedel I (1992) Neutralization of HIV-1 by anti-idiotypes to monoclonal anti-CD4: potential for idiotype immunization against HIV. J Immunol 149: 1452–1461

Syu W-J, Huang J-H, Essex M, Lee T-H (1990) The N-terminal region of the human immunodeficiency virus envelope glycoprotein contains potential binding sites for CD4. Proc Natl Acad Sci USA 87: 3695–3699

Takeda A, Tuazon CU, Ennis FA (1988) Antibody-enhanced infection by HIV-1 via Fc receptor-mediated entry. Science 242: 580–583

Tanabe TA, Tochikura TS, Blakeslee JJ, Olsen RG, Mathes LE (1992) Anti-human immunodeficiency virus (HIV) agents are also potent and selective inhibitors of feline immunodeficiency virus (FIV)-induced cytopathic effect: development of a new method for screening of anti-FIV substances in vitro. Antiviral Res 19: 161–172

Teshima G, Porter J, Yim k, Ling V, Guzzetta A (1991) Deamidation of soluble CD4 at asparagine-52 results in reduced binding capacity for the HIV-1 envelope glycoprotein gp120. Biochemistry 30: 3916–3922

Thali M, Olshevsky, U, Furman C, Gabuzda D, Li J, Sodroski J (1991) Effects in changes of gp120-CD4 binding affinity on human immunodeficiency type 1 envelope glycoprotein function and soluble CD4 sensitivity. J Virol 88: 5007–5012

Thieblemont N, HaeffnerCavaillon N, Ledur A, LageStehr J, ZieglerHeitbrock H, Kazatchkine MD (1993) CR1 (CD35) and CR3 (CD11b/CD18) mediate infection of human monocytes and monocytic cell lines with complement-opsonized HIV independently of CD4. Clin Exp Immunol 92: 106–113

Traunecker A, Luke W, Karjalainen K (1988) Soluble CD4 molecules neutralize human immunodeficiency virus type 1. Nature 331: 84–86

Truneh A, Buck D, Cassatt DR, Juszczak R, Kassis S, Ryu SE, Healey D, Swéet R, Sattentau Q (1991) A region in domain 1 of CD4 distinct form the primary gp120 binding site is involved in HIV infection and virus-mediated fusion. J Biol Chem 266: 5942–5948

Tsui P, Sweet RW, Sathe G, Rosenberg M (1992) An efficient phage plaque screen for the random mutational analysis of the interaction of HIV-1 gp120 with human CD4. J Biol Chem 267: 9361–9367

Valentin A, Lundin K, Patarroyo M, Asjo B (1990) The leukocyte adhesion glycoprotein CD18 participates in HIV-1-induced syncytia formation in monocytoid and T cells. J Immunol 144: 934–937

van den Berg L, Sadiq SA, Lederman S, Latov N (1992) The gp120 glycoprotein of HIV-1 binds to sulfatide and to the myelin associated glycoprotein. J Neurosci Res 33: 513–518

VermotDesroches C, Rigal D, Escaich S, Bernaud J, Pichoud C, Lamelin JP, Trepo C (1991) Functional epitope analysis of the human CD11a/CD18 molecule (LFA-1, lymphocyte function-associated antigen1) involved in HIV-1-induced syncytium formation. Scand Immunol 34: 461–470

Wang JH, Yan YW, Garrett TP, Liu JH, Rodgers DW, Garlick RL, Tarr GE, Husain Y, Reinherz EL, Harrison SC (1990) Atomic structure of a fragment of human CD4 containing two immunoglobulin-like domains (see comments). Nature 348: 411–418

Weiner DB, Huebner K, Williams WV, Greene MI (1991) Human genes other than CD4 facilitate HIV-1 infection of murine cells. Pathobiology 59: 361–371

Weiss R (1993) Cellular receptors and viral glycoproteins involved in retrovirus entry. In: Levy J (ed) The retroviridae, vol 2. Plenum, New York, pp 1–108

Willey RL, Martin MA, Peden K (1994) Increase in soluble CD4 binding to and CD4-induced dissociation of gp120 from virions correlates with infectivity of human immunodeficiency virus type 1. J Virol 68: 1029–1039

Wyatt R, Sullivan N, Thali M, Repke H, Ho D, Robinson J, Posner M, Sodroski J (1993) Functional and immunologic characterization of human immunodeficiency virus type 1 envelope glycoproteins containing deletions of the major variable regions. 67: 4557–4565

Zeira M, Byrn RA, Groopman JE (1990) Inhibition of serum-enhanced HIV-1 infection of U937 monocytoid cells by recombinant soluble CD4 and anti-CD4 monoclonal antibody. AIDS Res Hum Retroviruses 6: 629–639

# The Role of CD4 in HIV Envelope-Mediated Pathogenesis

R.F. Siliciano



## 1 Introduction

The high affinity interaction between the enve ope (env) glycoprotein of HIV-1 and the cell surface glycoprotein CD4 is a central molecular event in AIDS pathogenesis. This interaction not only provides the molecular basis for the tropism of HIV-1 for CD4⁺ cells, but, as is described in the present chapter, also represents a critical step in several complex pathogenic mechanisms that contribute to the development of immunodeficiency in HIV-1-infected individuals.

The env protein of HIV-1 is synthesized on the rough endoplasmic reticulum (ER) in productively infected cells as a glycosylated precursor protein, gp160. Newly synthesized gp160 molecules associate in the ER to form oligomeric structures that are probably tetramers (Earl et al. 1990). During transport through the Golgi, gp160 is cleaved by a cellular protease into two subunits: gp120, which has a high affinity binding site for CD4, and gp41, a transmembrane protein that is anchored in the plasma membrane of the infected cell. Cleavage is required for infectivity (McCune et al. 1988). The gp120 and gp41 subunits

Department of Medicine, 1049 Ross Building, Johns Hopkins University School of Medicine, 720 Rutland Avenue, Baltimore, MD 21205, USA remain noncovalently associated. Oligomers of gp120/gp41 complexes are incorporated into the viral envelope during budding where they form the characteristic glycoprotein "spikes." The high affinity interaction of this virion-associated env protein with CD4 dictates the trophism of the virus for cells expressing CD4 including CD4+ T lymphocytes and cells of the monocyte-macrophage lineage (DALGLEISH et al. 1984; KLATZMANN et al. 1984; MADDON et al. 1986; McDOUGAL et al. 1986a; LASKY et al. 1987; ASJO et al. 1987; LANDAU et al. 1988).

The binding of env protein to CD4 serves not only to attach virions to the surface of the CD4+ cell, but also to trigger conformational changes within the env protein complex. By analogy with other virus-cell fusion reactions, these conformational changes probably lead to exposure of the hydrophobic fusion domain of gp41 (WHITE et al. 1983; BULLOUGH et al. 1994). Insertion of this domain into the plasma membrane of the CD4+ cell initiates fusion of the viral envelope with the plasma membrane (WHITE et al. 1983), an event that gives the viral nucleocapsid access to the cytoplasm of the CD4+ cell. Thus, the association reaction between env protein and CD4 is the critical initial event in the infection of target cells by HIV-1. The biochemical characteristics of the HIV-1 env protein and its interaction with CD4 are discussed in more detail in the chapter by James et al. (this volume).

While the interaction between virion-associated gp120 and cell surface CD4 initiates the infection of susceptible cells, interactions between env protein and CD4 occur in several other situations in HIV-1 infection. For example, within infected cells, newly synthesized gp160 molecules interact with newly synthesized CD4 molecules in the ER (STEVENSON et al. 1988; CRISE et al. 1990). These intracellular gp160-CD4 interactions may result in toxicity. In addition, gp120/gp41 complexes expressed on the surface of an infected cell may interact with CD4 molecules on the surface of noninfected T lymphocytes present in the same microenvironment. The result in this case is a fusion reaction leading to the formation of a multinucleated giant cell or syncytium (SODROSKI et al. 1986; LIFSON et al. 1986a, b). Another interesting series of gp120-CD4 interactions may occur following the release or shedding of the gp120 subunit of the env protein from infected cells and from HIV-1 virions. The released gp120 retains the capacity to bind to CD4 and can therefore bind to CD4 molecules on noninfected CD4+ T cells, resulting in a number of important immunologic reactions. Interestingly, all of these types of env-CD4 interactions have been associated with pathogenic mechanisms in HIV-1 infection, and, as will be discussed below, may contribute to the inactivation and/or depletion of CD4+ T cells in infected individuals.

The depletion of CD4+ T cells is the central pathophysiologic feature of HIV-1 infection and is largely responsible for the profound immunodeficiency that is characteristic of late stages of the disease. CD4+ T cell depletion can be viewed in terms of a virus-induced alteration in cell kinetics within this compartment. In infected individuals, the rate of CD4+ T cell loss exceeds the rate at which CD4+ T cells are produced through thymic differentiation and/or clonal expansion of mature peripheral CD4+ T cells. CD4+ T cells may be lost through a number of potential mechanisms, only some of which operate on infected cells. Infected cells can be depleted through direct cytopathic effects of the virus or by virus-specific cytolytic T lymphocytes (CTLs) or antibody-dependent cell-mediated cytotoxicity (ADCC). Under some conditions, HIV-1 infection is highly cytopathic for CD4⁺ T cells. The direct cytopathic effects of the virus were initially considered to be responsible for the depletion of CD4⁺ T cells in HIV-1 infected individuals. However, only a very small proportion of the CD4⁺ T cells in seropositive individuals are productively infected at any given time (HARPER et al. 1986). This finding has given rise to a great deal of interest in other mechanisms by which HIV-1 might induce the destruction of CD4⁺ T cells, including CD4⁺ T cells that are not infected.

Noninfected CD4⁺ cells present in the vicinity of an infected cell may be killed through several mechanisms that involve gp120-CD4 interactions. Direct interactions between infected cells expressing gp120/gp41 and noninfected T cells expressing CD4 may lead to cell-cell fusion and the formation of syncytia (SODROSKI et al. 1986; LIFSON et al. 1986a, b). Inclusion of noninfected CD4⁺ T cells in short lived syncytia provides a mechanism for CD4⁺ T cell depletion. In addition, gp120 shed from infected cells can interact with CD4 on noninfected cells present in the same microenvironment. This in turn may render noninfected cells susceptible to destruction by CD16⁺ cytolytic effector cells in the presence of anti-gp120 antibody (LYERLY et al. 1987; TYLER et al. 1989; WEINHOLD et al. 1989) or by gp120-specific CD4⁺ CTLs (SILICIANO et al. 1988; LANZAVECCHIA et al. 1988; HAMMOND et al. 1992). The cross-linking of bound gp120 by anti-gp120 may prime CD4⁺ T cells for programmed cells death following antigen-driven activation (GROUX et al. 1992; BANDA et al. 1992). Finally, the binding of gp120 to CD4 on noninfected CD4⁺ T cells may suppress the function of these cells (SHALABY et al. 1987; KROWKA et al. 1988; MANN et al. 1987; CHIRMULE et al. 1988, 1990; WEINHOLD et al. 1989; HABESHAW et al. 1990; MANCA et al. 1990; MARTIN et al. 1994). Each of these mechanisms will be discussed in detail below.

## 2 Intracellular Interactions Between gp160 and CD4

In infected cells, newly synthesized env protein can interact intracellularly with CD4 produced in the same cell. This interaction results in toxicity that may in part explain the ability of HIV-1 to kill infected CD4⁺ lymphoblasts. Two general types of HIV-1- induced cytopathic effects have been observed in vitro. The first is syncytium formation, which is discussed in the next section. The second type of cytopathic effect operates at the level of individually infected cells; that is, under some conditions, HIV-1-infected T cells appear to die from the infection independent of any cell-cell fusion events (SOMASUNDARAN and ROBINSON 1987; STEVENSON et al. 1988). It was initially considered that the accumulation of unintegrated viral DNA might be responsible for this type of cytopathic effect. High levels of unintegrated viral DNA produced by reverse transcription of RNA molecules from newly infecting virions have been associated with cytopathic effects in in vitro infections by avian and feline retroviruses. However, several recent studies have demonstrated a dissociation between the accumulation of unintegrated HIV-1 DNA and single cell killing of HIV-1-infected cells (BERGERON and SODROSKI 1992; LAURENT-CRAWFORD and HOVANESSIAN 1993). This has led to increased interest in another major theory regarding the mechanism of single cell killing by HIV-1, one that involves the formation of intracellular complexes between gp160 and CD4.

In productively infected $CD4^+$ lymphoblasts, both gp160 and CD4 are synthesized on the rough ER. Therefore, the possibility for intracellular interaction of newly synthesized gp160 and CD4 molecules exists. Rose and colleagues have demonstrated that gp160-CD4 complexes form in the ER and are retained there (CRISE et al. 1990). Retention appears to result from the inefficient transport of gp160 itself (WILLEY et al. 1988). This phenomenon provides a partial explanation for early observations that productively infected cells had reduced surface expression of CD4 (DALGLEISH et al. 1984; KLATZMANN et al. 1984; HOXIE et al. 1986). More recent studies (CRISE and ROSE 1992) indicate that the gp160-CD4 complexes that form in the ER also include the protein tyrosine kinase $p56^{lck}$, which is associated with the cytoplasmic domain of CD4 (see the chapter by Marsh and Pelchen-Mathews). It has been proposed that mislocalization of this kinase to the cytoplasmic face of the ER may have some role in viral cytopathic effects (CRISE and ROSE 1992).

Compelling evidence that HIV-1-induced single cell killing involves intracellular interactions between the env protein and CD4 comes from in vitro studies showing that the susceptibility of cells to the cytopathic effects of HIV-1 infection is related to the level of CD4 expression. Following infection of susceptible $CD4^+$ T cell lines with HIV-1, cells that grow out with stably integrated provirus tend to have low or absent CD4 expression, reflecting a selection against coexpression of env protein and CD4 (HOXIE et al. 1986). In addition, cells of the monocyte/macrophage lineage, which express only low levels of CD4, can be infected by HIV-1 but do not show cytopathic effects (GARTNER et al. 1986; HO et al. 1986; NICOLSON et al. 1986; SALAHUDDIN et al. 1986). The level of expression of CD4 by subclones of the monocytic line U937 is correlated with susceptibility of the clones to HIV-1-induced cytopathic effects (ASJO et al. 1987).

In an interesting series of studies, Nomoto and colleagues have analyzed the cytopathic effects resulting from *env* gene expression (KOGA et al. 1990a, b). $CD4^+$ and $CD4^-$ sublines of U937 and of the T cell tumor line Jurkat were transfected with a vector carrying the HIV-1 *env* gene under the control of the inducible metallothionein promoter. Cytopathic effects were observed upon induction of *env* gene expression only in the $CD4^+$ sublines, suggesting that *env* gene expression is not in itself toxic to the cell. Rather it is the coexpression of gp160 and CD4 that produces cytopathic effects at the single cell level. Coexpression resulted in the formation of intracellular complexes between gp160 and CD4. Coexpression of the gp120 subunit of the env protein and CD4 did not produce cell death (KOGA et al. 1994). Cell death induced after gp160-CD4 coexpression appeared to involve an apoptotic process (LU et al. 1994), as is the case with single cell killing of HIV-1-infected cells (TERAI et al. 1991; LAURENT-CRAWFORD et al. 1991; MARTIN et al. 1994). In addition, in CD4+ U937 cells, accumulations of env protein at the nuclear pores were noted, leading to the proposal that the gp160-CD4 complexes might accumulate at the nuclear pores and produce cytopathic effects by interfering with the transport of molecules between the nucleus and the cytoplasm. Nuclear transport of a microinjected substrate was shown to be abnormal in transfected cells in which gp160-CD4 complexes formed (KOGA et al. 1991). Whether such complexes localize to nuclear pores in HIV-1-infected T cells was not determined. These studies demonstrate that in CD4+ cells cytopathic effects can result from the expression of the HIV-1 *env* gene in the absence of other viral genes (except *rev* which is necessary for *env* expression) and that formation of gp160-CD4 complexes may be involved in inducing these cytopathic effects, possibly through alterations in nuclear transport.

In another interesting study, the env protein has been implicated in HIV-1-induced cytopathic effects at the single cell level. STEVENSON and colleagues (1990) carried out successive infections of the cytolysis-sensitive MT-4 T cell line and isolated surviving cells from each round of infection. This approach led to the amplification of an HIV-1 variant with a normal capacity for replication and syncytium formation but a reduced ability to mediate single cell killing. The critical differences responsible for the loss of cytopathic potential were mapped to the *env* gene. Two amino acid substitutions were present in the gp120 subunit, each of which produced the loss of a conserved glycosylation site. In another study, KOWALSKI and colleagues (1991) have shown that mutations in the $NH_2$-terminal of gp41 lead to a loss of syncytium formation and single cell killing without affecting viral replication. Thus subtle differences in the env protein can affect cytopathic potential. It is also important to note that the naturally occurring variability in HIV-1 has also been shown to affect cytopathic potential. As is discussed below, not all HIV-1 isolates induce cytopathic effects in infected cells.

In summary, in cells coexpressing env protein and CD4, intracellular complexes form in the ER and tend to remain there. These complexes may induce cytopathic effects. There is evidence that import of proteins into the nucleus is abnormal in cells containing these complexes. Another possibility is that the interaction of newly synthesized gp160 and CD4 molecules induces close apposition of opposing faces of the ER membranes, thereby disrupting some aspect of normal ER function.

It should be pointed out that the formation of intracellular complexes between gp160 and CD4 can in principle limit the transport, cleavage, and surface expression of env protein. Recent work by WILLEY and colleagues (1992a,b) demonstrates that the vpu protein of HIV-1 functions to destabilize intracellular gp160-CD4 complexes, thereby increasing the processing of the env protein. In addition, vpu promotes the degradation of CD4. Thus vpu, which is encoded on the same bicistronic mRNA as env (SCHWARTZ et al. 1990), may prolong the survival of infected cells by reducing the cytopathic effects resulting from the formation of intracellular complexes between gp160 and CD4. In addition, the enhanced processing and surface expression of env protein may promote the release of infectious virus. The HIV-1 nef protein also appears to affect CD4 expression by infected cells. As is discussed by Marsh and Pelchen-Matthews in this volume, the nef protein plays a role in down-regulation of cell surface CD4.

## 3 Syncytium Formation

In early studies of HIV-1 infection of susceptible cell lines in vitro, the formation of multinucleated giant cells was noted. Subsequent studies showed that infected cells expressing surface env protein could fuse with noninfected cells expressing CD4 (Sodroski et al. 1986; Lifson et al. 1986a, b). Inclusion of noninfected $CD4^+$ T cells in short lived syncytia therefore provides a potential mechanism for $CD4^+$ T cell depletion. The process of syncytium formation involves membrane fusion events initiated by the interaction of gp120 and CD4 on opposing cell surfaces. The fusion reaction is likely to be very similar to that which occurs upon binding of HIV-1 virions to $CD4^+$ cells. However, cell adhesion molecules appear to play a role in inducing the close apposition of adjacent cells that is necessary for sycytium formation. This was first demonstrated by Hildreth and colleagues who showed that antibodies to the cell adhesion molecule LFA-1 blocked syncytium formation (Hildreth and Orentas 1989).

The extent to which syncytium formation contributes to $CD4^+$ T cell depletion in vivo is unclear. In pathologic studies of lymph nodes and spleen from infected individuals, giant cells are observed only rarely (Macher et al. 1988). In addition, not all HIV-1 isolates induce syncytia in vitro. Isolates from different infected individuals show variation in biological properties including rates of replication in vitro, ability to infect various T cell and monocyte-macrophage cell lines, ability to induce syncytia, and ability to induce cytopathic effects (Fisher et al. 1988; Cheng-Mayer et al. 1988; Fenyo et al. 1988). Sequence variation in the *env* gene affects the capacity of HIV-1 strains to induce syncytium formation and syncytium-independent cytolysis (York-Higgins et al. 1990; Stevenson et al. 1990; Kowalski et al. 1991). In some infected individuals, the capacity of HIV-1 isolates to induce cytopathic effects in vitro increases with disease progression. Levy and colleagues showed initially, in a small series of seropositive individuals, that isolates obtained before the onset of symptoms were not cytopathic whereas isolates taken subsequent to the onset of symptoms were able to induce either syncytia or balloon degeneration in infected cultures (Cheng-Mayer et al. 1988). Other studies have shown that isolates with the capacity to induce syncytia emerge with disease progression in certain individuals. However, in about 50% of cases, progression to AIDS is not accompanied by a transition from the nonsyncytia-inducing (NSI) to the syncytia-inducing (SI) phenotype (Tersmette et al. 1989). High in vitro replication rates appear to correlate with the SI phenotype and with rapid $CD4^+$ T cell loss in vivo. Several recent studies have confirmed association of the SI phenotype with a more rapid decline in CD4+ T cell counts and progression to AIDS (BOZZETTE et al. 1993; KOOT et al. 1993; RICHMAN and BOZZETTE 1994). Nevertheless, CD4+ T cell depletion occurs during the asymptomatic phase of infection when the predominant viral species are usually of the NSI phenotype. Thus, the precise role of syncytium formation in the pathogenesis of immunodeficiency in HIV-1 infection remains to be clarified.

## 4 Shedding of gp120

Several other mechanisms for the destruction or inactivation of noninfected CD4+ T cells depend upon the release of gp120 from the surface of infected cells or HIV-1 virions and the subsequent binding of shed gp120 to CD4 molecules on noninfected CD4+ T cells. As a prelude to the discussion of these mechanisms, the evidence that shedding of gp120 occurs to a significant extent in vivo will be reviewed.

The association between gp120 and gp41 is noncovalent, and spontaneous dissociation can occur. Thus gp120 is shed from infected cells and from HIV-1 virions (SCHNEIDER et al. 1986; GELDERBLOM et al. 1985, 1987; MCKEATING et al. 1991). The shedding of subunits of retroviral env proteins was originally observed for murine leukemia viruses (BOLOGNESI et al. 1975). Early studies on cultured HIV-1 demonstrated that the glycoprotein spikes on the surface of the virion were labile, the lability being associated with loss of the gp120 subunit (GELDERBLOM et al. 1985, 1987). As a result, supernatants of HIV-1-infected cells contain free gp120 in addition to virion-associated gp120 (SCHNEIDER et al. 1986; MCKEATING et al. 1991). The rate of gp120 shedding varies among HIV-1 isolates (MCKEATING et al. 1991). The half-time for the dissociation of gp120 from virions of the LAI isolate of HIV-1 is 30 h at 37°C (LAYNE et al. 1992). As is discussed in Chap. 7, the dissociation of gp120 from gp41 is greatly accelerated by the binding of soluble CD4 (MOORE et al. 1990).

Critical to evaluating the potential in vivo role of any of the potentially deleterious reactions described below is the question whether the amount of gp120 released in the local microenvironment is sufficient to induce the relevant effects. Simple mass action principles suggest that uptake of shed gp120 may occur most readily in the lymph nodes because of the extremely high cell concentrations. Another very important variable is the concentration of anti-gp120 antibodies. Uptake of gp120 can be readily blocked by some anti-gp120 antibodies, those that block the CD4 binding site. However, only a small fraction of the anti-gp120 antibody response in infected individuals appears to be directed at the CD4 binding site (MCDOUGAL et al. 1986b; SCHNITTMAN et al. 1988; THALI et al. 1991). These antibodies have been associated with the broadly cross-reactive, conformation-sensitive, late-arising neutralizing activity detected in sera from infected individuals. While antibodies to the CD4-binding site on gp120 may block the uptake of shed gp120 by $CD4^+$ lymphocytes, it is also important to consider that the likelihood of such blocking is reduced in the lymph nodes where the concentration of target cells is several orders of magnitude higher than those used in most in vitro studies.

To date only a small number of studies have addressed the issue of whether uptake of shed gp120 occurs in vivo. OH and colleagues (1992) have used a sandwich ELISA technique to detect HIV-1 env protein in the serum of 22/32 AIDS patients. A modified assay detected env-containing immune complexes in four out of four patients with AIDS-related complex (ARC) and 11/11 patients with AIDS. With respect to the critical question of whether or not shed gp120 binds to uninfected $CD4^+$ T cells in sufficient amounts to render the cells subject to the destructive mechanisms described below, there is little available data. AMADORI et al. (1992) have demonstrated the masking of some CD4 epitopes on peripheral blood T cells from patients with ARC or AIDS. These investigators have suggested that this is due to the binding of gp120 and anti-gp120. Unfortunately, it was not possible in this study to prove that gp120 was present on the cells although the authors do provide evidence that some of the CD4-associated antibody is gp120-specific. Another group has used a relatively simple flow cytometry assay to detect gp120 on $CD4^+$ T cells from infected individuals and has shown that the fraction of cells bearing gp120 increases as $CD4^+$ T cell counts decline (DANIEL et al. 1993a, b). Immunoglobulin and the complement fragment C3dg were also detected on $CD4^+$ T cells in these studies. While this work provides suggestive evidence that shedding and CD4-dependent uptake of gp120 can occur in vivo in infected individuals, much additional work will be required to determine the functional significance of this transfer of gp120.

## 5 gp120-Specific Antibody-Dependent Cell-Mediated Cytotoxicity

The results discussed above indicate that gp120 is shed from the surfaces of infected cells and from HIV-1 virions and may then associate in vivo with CD4 molecules on the surfaces of noninfected $CD4^+$ T cells. It is therefore important to consider what consequences this shedding and uptake of viral env protein might have. One very interesting consequence is that noninfected $CD4^+$ T cells that have taken up gp120 may become susceptible to lysis by an ADCC mechanism (Fig. 1). This is an important effector mechanism that normally operates on host cells expressing viral proteins on the cell surface (see BRENNER et al. 1990 for a review of ADCC in HIV-1 infection). In an important early study, WEINHOLD and colleagues showed that noninfected $CD4^+$ T cells that had been incubated with purified gp120 could be destroyed through an ADCC mechanism that was dependent on the presence of anti-gp120 antibodies (LYERLY et al. 1987). Anti-gp120 antibodies capable of mediating ADCC are present in the vast

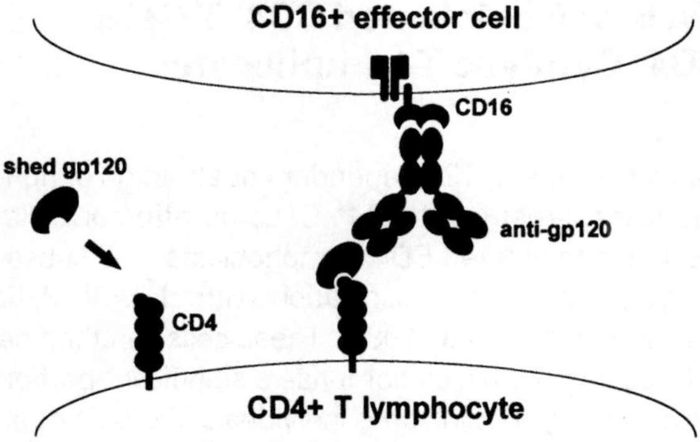


**Fig. 1.** Lysis of noninfected $CD4^+$ T lymphocytes by a gp120-specific, antibody-dependent cell-mediated cytotoxicity (ADCC) mechanism. Shed HIV-1 gp120 binds with high affinity to CD4 on noninfected $CD4^+$ T cells. In the presence of anti-gp120 antibodies, these cells can be recognized and lysed by $CD16^+$ effector cells. This mechanism can lead to the destruction of noninfected $CD4^+$ T cells that are in the resting or activated state

majority of infected individuals. These antibodies can bind to cell-associated gp120 and trigger activation of a subpopulation of non-B, non-T lymphocytes bearing the type III Fcγ receptors (CD16). Upon interaction with antibody-bearing target cells, $CD16^+$ lymphocytes mediate target cell lysis through a mechanism that resembles target cell destruction by $CD8^+$ CTLs. WEINHOLD and coworkers (1989) observed lysis in vitro at gp120 concentrations as low as $5 \times 10^{-10}$ *M*. Saturation of surface CD4 was not required for lysis; rather lysis could be detected under conditions in which only a very small fraction of the surface CD4 molecules had bound gp120. Bound gp120 was gradually lost from the cell surface over 12 h by dissociation and/or internalization, providing sufficient time for these cytolytic reactions. In principle, this mechanism can operate on both resting and activated $CD4^+$ T cells. This group has also shown that circulating $CD16^+$ effector cells from infected individuals carry sufficient amounts of bound cytophilic anti-gp120 antibodies to mediate the in vitro lysis of gp120-coated $CD4^+$ targets cells in the absence of additional anti-env antibodies (TYLER et al. 1989).

Taken together, these observations raise the possibility that noninfected $CD4^+$ lymphocytes present in the vicinity of productively infected cells may absorb sufficient amounts of shed gp120 to become susceptible to destruction by $CD16^+$ effector cells. As is the case with other proposed mechanisms for $CD4^+$ T cell depletion, it is not yet clear to what extent this mechanism contributes to $CD4^+$ T cell depletion in vivo. In addition to the rate of gp120 shedding and the concentration of blocking anti-gp120 antibodies, another important factor to be considered in evaluating the role of this pathway in vivo is the progressive impairment of $CD16^+$ effector cells that occurs at later stages of HIV-1 infection (BRENNER et al. 1990; TYLER et al. 1990).

## 6 gp120-Dependent Lysis of Noninfected CD4⁺ T Cells by gp120-Specific CD4⁺ Cytolytic T Lymphocytes

An even more complex mechanism for the gp120-dependent destruction of noninfected $CD4^+$ T cells involves class II-restricted $CD4^+$ CTLs as effector cells. When shed gp120 binds to CD4 on noninfected $CD4^+$ lymphoblasts, it is subsequently processed within the lymphoblasts for association with class II MHC molecules (SILICIANO et al. 1988; LANZAVECCHIA et al. 1988). These cells can then be lysed by gp120-specific $CD4^+$ CTLs (Fig. 2), which comprise a significant portion of the normal human T cell response to gp120 in some individuals (SILICIANO et al. 1988; ORENTAS et al. 1990; STANHOPE et al. 1993a, b).

The elucidation of this novel pathway for the depletion of noninfected $CD4^+$ T cells was dependent upon the discovery that T cells can process protein antigens that have bound with high affinity to a T cell surface structure (SILICIANO et al. 1988; LANZAVECCHIA et al. 1988). Although activated human T cells are uniformly positive for expression of class II MHC gene products, no function has been ascribed to the class II molecules expressed by T cells. It appears unlikely that T cells play a general role in the presentation of exogenous antigens because they lack an efficient mechanism for antigen uptake. In fact, processing of exogenous antigens by T cells has only been demonstrated in specialized situations, in which the T cells have a high affinity uptake mechanism for a particular exogenous antigen (SILICIANO et al. 1988; LANZAVECCHIA et al. 1988). The processing of HIV-1 gp120 represents such a situation. In this case, processing is strictly dependent upon high affinity uptake via CD4. Presentation of gp120 by T cells can be blocked by anti-CD4 antibodies that block the gp120 binding site or by soluble CD4 (SILICIANO et al. 1988). Interestingly, monocytes can also take up gp120 via CD4 (SILICIANO

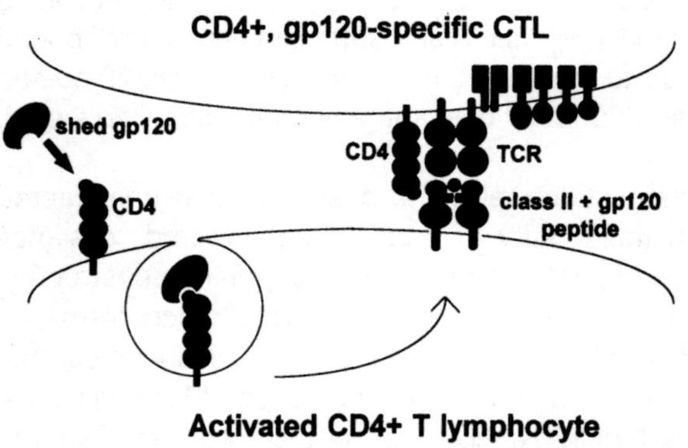


**Fig. 2.** Lysis of noninfected $CD4^+$ T lymphoblasts by gp120-specific $CD4^+$ cytolytic T lymphocytes (CTLs). Shed HIV-1 gp120 binds with high affinity to CD4 on noninfected $CD4^+$ T cells. The shed gp120 is subsequently internalized and processed within the lymphoblasts for association with class II MHC molecules. Class II molecules carrying peptides derived from gp120 are expressed on the cell surface. These cells can then be lysed by $CD4^+$, MHC class II-restricted CTL specific for gp120. This mechanism operates only on activated $CD4^+$ T cells. Resting T cells do not express class II molecules

et al. 1989). This specific uptake reaction lowers by 30-fold the concentration of gp120 needed to trigger a comparable monocyte-dependent gp120-specific T cell proliferative response.

The effector cells that mediate the destruction of gp120-pulsed $CD4^+$ lymphoblasts in the mechanism described above are class II-restricted $CD4^+$ CTLs (SILICIANO et al. 1988). Conventional $CD8^+$ CTLs specific for gp120 are unable to lyse gp120-pulsed autologous lymphoblasts even under conditions in which the same targets can be lysed by gp120-specific $CD4^+$ CTLs (HAMMOND et al. 1992). This reflects the fact that exogenous antigens do not generally have access to the class I-restricted antigen processing pathway (MORRISON et al. 1986; MOORE et al. 1988). The gp120-dependent destruction of noninfected $CD4^+$ lymphoblasts was originally demonstrated using gp120-specific $CD4^+$ CTL clones isolated from naive human donors (SILICIANO et al. 1988). More recently, the same phenomenon has been demonstrated for env-specific $CD4^+$ CTLs isolated from seronegative participants in trials of several different purified recombinant env protein vaccines (ORENTAS et al. 1990; HAMMOND et al. 1992). $CD4^+$ CTLs specific for the env protein have been detected in a significant fraction of the recipients of all env-based vaccines tested to date in humans, including various recombinant env protein vaccines (ORENTAS et al. 1990; STANHOPE et al. 1993a, b) and recombinant vaccinia or avipox virus vaccines carrying the HIV-1 *env* gene (HAMMOND et al. 1992;M. Egan and R.F. Siliciano 1994, unpublished observations). Significantly, because endogenously synthesized env protein is processed for association with class II MHC gene products (POLYDEFKIS et al. 1990; CALLAHAN et al. 1993), HIV-1-infected cells are also readily lysed by these vaccine-induced $CD4^+$ CTLs (ORENTAS et al. 1990). The mechanism of cytolysis by $CD4^+$, env-specific CTLs resembles that utilized by $CD8^+$ CTLs in that lysis is rapid, $Ca^{2+}$-dependent, and highly specific, and does not require de novo RNA or protein synthesis by either the effector or target cell. The lytic protein perforin is detectable in a granular distribution in human $CD4^+$ CTLs (MISKOVSKY et al. 1994). Recognition of env antigen by these $CD4^+$ CTLs shows a high degree of antigen specificity and precise MHC restriction (SILICIANO et al. 1988; HAMMOND et al. 1991; JOHNSON et al. 1994) and is not due to formation of syncytia between $CD4^+$ effector cells and env-expressing target cells (MISKOVSKY et al. 1994).

Although the results cited above demonstrate that CTLs from the $CD4^+$ T cell subset constitute part of the normal human T cell response to gp120, it is less clear that the gp120-specific $CD4^+$ CTLs represent an important effector population in infected individuals. Env-specific $CD4^+$ CTL responses have been detected in seropositive individuals (CURIEL et al. 1993; KUNDU and MERIGAN 1992), and they can be enhanced by immunization with recombinant HIV-1 env protein (KUNDU et al. 1992). However, most of the MHC-restricted env-specific cytolytic activity in peripheral blood of infected individuals is mediated by $CD8^+$ CTLs (see WALKER and PLATA 1990, for a review). $CD4^+$ CTL clones specific for HIV-1 gag protein have also been isolated from HIV-1 seropositive individuals (LITTAUA et al. 1992). One of the problems in determining whether gp120-specific $CD4^+$ CTLs are present in infected individuals is that these cells are typically isolated by in vitro stimulation with recombinant env protein derived from a standard laboratory strain of HIV-1. Yet the level of interisolate sequence variability in gp120 is so great that most CD4⁺ T cell clones specific for gp120 exhibit a highly type-specific reactivity pattern (SILICIANO et al. 1988; CALLAHAN et al. 1990; ORENTAS et al. 1990; HAMMOND et al. 1992). Therefore, understanding the nature of the CD4⁺ T cell response to gp120 in infected individuals may require cloning and expression of *env* genes from HIV-1-infected individuals, followed by an analysis of autologous CD4⁺ T cell responses to the relevant expressed env proteins. Another problem is that gp120 itself is immunosuppressive (see below), further complicating the analysis of the gp120-specific CD4⁺ T cell response. Therefore, at the present time, it is unclear whether gp120-specific CD4⁺ CTLs are present in infected individuals at frequencies high enough to make this mechanism for CD4⁺ depletion quantitatively significant. One special situation in which gp120-specific CD4⁺ CTLs might contribute to CD4⁺ T cell depletion is the case of vaccine failures. Because env-based vaccines clearly induce gp120-specific CD4⁺ CTLs in some recipients, these cells may prove deleterious in immunized individuals who become infected.

Another critical issue is whether the degree of shedding of gp120 in vivo allows uptake by CD4⁺ lymphoblasts of sufficient amounts of gp120 to render these cells susceptible to lysis of CD4⁺ CTLs. In this regard, it should be noted that the autocytolytic mechanism described above can operate even when the concentration of gp120 is below the $k_d$ of the gp120-CD4 interaction, indicating that uptake and processing of only a small number of gp120 molecules may be sufficient to render a T cell susceptible to lysis. The gp120-specific lysis of non-infected CD4⁺ lymphoblasts by gp120-specific CD4⁺ CTLs can be demonstrated in in vitro cultures in which the only source of gp120 is that shed from infected cells (HAMMOND et al. 1992). Recent results from our laboratory suggest that the processing of gp120 by T cells is highly dependent upon the state of activation of the target T cell and occurs efficiently only in highly activated cells (M. Polydefkis and R. Siliciano, unpublished results). In addition, this pathway can be inhibited by anti-gp120 antibodies that are present in the sera of infected individuals. The in vivo significance of this CD4⁺ CTL-mediated pathway for CD4⁺ T cell destruction relative to other proposed pathways for CD4⁺ T cell destruction such as lytic infection and syncytium formation remains to be established.

## 7 gp120-Dependent Priming for Programmed Cell Death

A final mechanism for CD4⁺ T cell depletion also involves the binding of shed gp120 to noninfected T cells. Recent in vitro studies by BANDA et al. (1992) have shown that the binding of free recombinant gp120 to CD4 followed by cross-linking of bound gp120 with anti-gp120 antibodies primes cells for programmed cell death (apoptosis) upon subsequent activation through the T cell receptor (Fig. 3). Anti-CD4 antibodies alone were previously shown to have a similar effect when cross-linked on the surface of murine CD4⁺ T cells with anti-immunoglobulin (NEWELL et al. 1990). Consistent with these findings is evidence that T cells from infected individuals appear to be primed in vivo for programmed cell death. GROUX et al. (1992) have shown that in vitro stimulation of peripheral blood mononuclear cells (PBMCs) from healthy seropositive donors with a bacterial superantigen or with certain mitogens results in programmed cell death in the responding CD4⁺ T cell population. This effect did not appear to be due to HIV-1 infection of the responding cells in culture as it was not blocked by AZT. Rather, these workers suggest that CD4⁺ T cells from infected individuals may be primed in vivo for activation-induced programmed cell death. It is not yet clear whether priming occurs by the gp120-dependent mechanism described above or by another mechanism. Interestingly, antibodies to the T cell surface protein CD28 blocked programmed cell death and restored proliferative responses to superantigens and to recall antigens, suggesting that costimulatory signals could interfere with this pathway (GROUX et al. 1992). In general, activation is necessary for cell death in these models and might be induced in vivo by conventional antigens. In this case, the cells most frequently activated would be depleted quickly. In principle, T cells could also be activated by a putative HIV-1-encoded or HIV-1-induced superantigen (IMBERTI et al. 1991; LAURENCE et al. 1992). However, several recent attempts to detect the selective expansion or deletion of subsets of T cells expressing particular T cell receptor Vβ genes have given negative results, indicating that HIV-1 probably does not encode a conventional superantigen (BOLDT-HOULE et al. 1993; BOYER et al. 1993; POSNETT et al. 1993; CHEN et al. 1993).

While cross-linking of gp120 bound to CD4 provides one plausible mechanism for activation-induced apoptosis, there are problems with this model. Some groups have failed to detect apoptosis following treatment of human CD4⁺ T cells with gp120 and anti-gp 120 (LIEGLER and STITES 1994; MARTIN et al. 1994). The

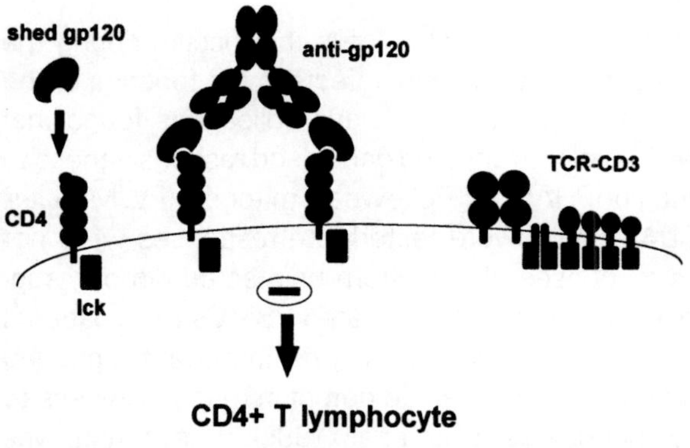


**Fig. 3.** Apoptosis or inactivation of noninfected CD4⁺ T cells by cross-linking of CD4 with gp120 and anti-gp120. The binding of shed gp120 to CD4 followed by cross-linking of bound gp120 with anti-gp120 antibodies results in signaling events that may prime noninfected CD4⁺ T cells for programmed cell death or apoptosis upon subsequent antigen-driven activation through the TCR. Alternatively, cross-linking of CD4 may lead to functional inactivation of the responding cells

discrepancy may reflect differences in the methods used to activate T cells in the different studies. In particular, stimulations carried out with anti-CD3 or anti-T cell receptor antibodies in the absence of cells capable of providing costimulatory signals may be more likely to induce apoptosis. Although apoptosis has not been observed in some studies, a suppression of $CD4^+$ T cell function has been noted consistently, as is discussed in the next section. Again, it is also unclear whether gp120 is shed in sufficient amounts in vivo to prime cells for apoptosis and whether uptake can occur in the presence of antibodies of the CD4 binding site on gp120 (see Sect. 4). Most studies to date have used recombinant gp120, monoclonal anti-gp120 antibodies, and nonspecific polyclonal activation stimuli rather than gp120 shed from infected cells, antisera from infected individuals, and antigen-specific stimulation. There is also evidence that additional mechanisms for T cell apoptosis may be involved. Some groups have reported that both $CD4^+$ and $CD8^+$ T cell populations from infected individuals show an increased frequency of cells undergoing spontaneous and activation-induced apoptosis in culture (MEYAARD et al. 1992; GOUGEON et al. 1993). Priming of $CD8^+$ cells for programmed cell death is most likely *not* due to a gp120-dependent mechanism. The priming of $CD4^+$ and $CD8^+$ T cells for programmed cell death can be detected at all stages of HIV-1 infection, including acute infection, but does not correlate with CD4 counts or the appearance of SI-type virus (MEYAARD et al. 1994). Interestingly, T cells from HIV-1-infected chimpanzees do not undergo activation-induced apoptosis (GOUGEON et al. 1993). Thus it remains an open question whether physiological activation by antigen (or superantigen) in vivo results in apoptosis of $CD4^+$ T cells primed by gp120-CD4 interactions.

## 8 gp120-Induced Suppression of T Cell Responses

In addition to the quantitative depletion of $CD4^+$ T cells that occurs during the course of HIV-1 infection, there are also qualitative defects in the function of the surviving $CD4^+$ T cells.In a pivotal early study, LANE and colleagues found that unfractionated PBMCs from AIDS patients showed decreased responsiveness to the mitogens Phytohemagglutinin (PHA) and Pokeweed mitogen (PWM) (LANE et al. 1985). When purified $CD4^+$ T cells were tested, the responses were not significantly different than the responses of cells from uninfected donors, suggesting that the lower response seen with unfractionated PBMCs from patients was simply due to a lower percentage of $CD4^+$ T cells. Most importantly, purified $CD4^+$ T cells from AIDS patients did show a marked defect in responsiveness to soluble antigen presented by monocytes (LANE et al. 1985). This defect was shown to be intrinsic to the T cell and did not occur at the level of the antigen-presenting cell (KOENIG and FAUCI 1988).

There are conflicting studies dealing with the question of when during the course of HIV-1 infection this functional defect becomes manifest. In some studies, T cell proliferative responses to soluble recall antigens are spared until immediately before the onset of AIDS (GIORGI et al. 1987; GURLEY et al. 1989). However, SHEARER et al. (1986) have shown that the $CD4^+$ T cell help needed for the in vitro generation of anti-influenza CTL responses is defective in some healthy HIV-1-seropositive individuals. Healthy seropositive individuals also show significantly decreased proliferative responses to immobilized anti-CD3 antibodies. This has been attributed to selective depletion of $CD4^+$ T cells with a memory phenotype that normally account for much of this response (VAN NOESEL et al. 1994).

The mechanism for the decreased responsiveness of $CD4^+$ T cells from infected individuals to soluble antigens is of great interest. Although HIV-1-infected $CD4^+$ T cells do not respond normally to stimulation through the antigen receptor for a variety of reasons, infected cells represent only a small fraction of the total $CD4^+$ T cell population. It is therefore unclear why T cell populations from infected individuals respond so poorly to antigen stimulation. One possibility is that the functional defects in surviving $CD4^+$ T cells actually reflect the fact that antigen-specific memory cells are preferentially deleted resulting in low responses to standard recall antigens. Other theories involve the potential immunosuppressive effects of HIV-1 proteins released from infected cells. The best studied case of immunosuppression by an HIV-1 gene product is that of the effect of HIV-1 gp120 on $CD4^+$ T cell responses. A large number of studies documenting a suppressive effect of gp120 on $CD4^+$ T cell responses have been published (SHALABY et al. 1987; KROWKA et al. 1988; MANN et al. 1987; CHIRMULE et al. 1988, 1990; WEINHOLD et al. 1989; HABESHAW et al. 1990; MANCA et al. 1990; MARTIN et al. 1994). The most detailed studies have been carried out on antigen-specific T cell clones. For example, CHIRMULE et al. (1990) have shown that purified gp120 can inhibit the proliferative responses of a $CD4^+$, tetanus toxoid-specific T cell clone at doses below $1 \times 10^{-10}$ *M*. In this study, early events in signal transduction were shown to be inhibited by gp120.

One explanation for the inhibitory effect of gp120 is that it prevents the interaction of CD4 with its ligand on the target cell, MHC class II. The interaction of $CD4^+$ T cells with cell-size lipid vesicles bearing class II MHC molecules can be inhibited by gp120, although higher concentrations are required ($4 \times 10^{-8}$ *M*) (ROSENSTEIN et al. 1994). Similarly, the interaction of CD4-transfected COS cells with class II bearing B cells can be completely inhibited by $2 \times 10^{-7}$ *M* gp120 (CLAYTON et al. 1989). These results are consistent with mutagenesis studies indicating that the region of CD4 that interacts with gp120 also appears to be involved in interactions with MHC class II. The class II binding site on CD4 appears to be larger and more complex than the gp120 binding site (see Chap. 2 for a comprehensive discussion of the relationship between the gp120 and class II binding sites on CD4).

Another possibility is that the binding of gp120 to CD4 transmits signals which interfere with activation through the T cell antigen receptor, as was originally observed with anti-CD4 antibodies (BANK and CHESS 1985). CENTER and colleagues have reported that treatment of resting human T cells with partially purified gp120 increases intracellular levels of inositol trisphosphate and $Ca^{2+}$, induces expression of the interleukin-2 (IL-2) receptor, and initiates a chemotactic response (KORNFELD et al. 1988). In other studies, increases in intracellular $Ca^{2+}$ have not been noted following binding of gp120 (MITTLER and HOFFMANN 1989; HORAK et al. 1990) or HIV-1 virions (HORAK et al. 1990). Interestingly, binding of gp120 and then anti-gp120 to CD4⁺ peripheral blood T cells can inhibit the increases in intracellular $Ca^{2+}$ seen after T cell receptor cross-linking (HORAK et al. 1990). The effect was observed at gp120 concentrations as low as $1 \times 10^{-10}$ *M* and with sera from 14 of 14 AIDS patients. A number of more recent studies have confirmed that gp120 in combination with anti-gp120 can suppress $CD4^+$ T cell responses (LIEGLER and STITES 1994; MARTIN et al. 1994), but it remains unclear whether this effect is primarily due to the blocking of interactions with class II MHC or to intracellular signaling effects.

## 9 Summary

There is now compelling evidence that env-CD4 interactions are central to several complex pathogenic mechanisms in HIV-1 infection. In addition to mediating virus attachment to $CD4^+$ cells, the high affinity interaction of env protein with CD4 is also important in initiating both syncytium formation and syncytium-independent cytopathic effects. In addition, shed gp120 can bind to CD4 on noninfected cells and interfere with the function of these cells while at the same time rendering the cells susceptible to destruction by ADCC, by $CD4^+$ CTLs or by programmed cell death induced by cross-linking of CD4 with gp120 and anti-gp120 followed by cellular activation. Although all of these mechanisms have been demonstrated to operate in vitro, it remains unclear how important each mechanism is in vivo. Nevertheless, the central role of env-CD4 interactions in all of these pathogenic mechanisms highlights the importance of developing effective low molecular weight inhibitors of this reaction.

## References

Amadori A, de Silvestro G, Zamarchi R, Veronese ML, Massa MR, Schiavo G, Panozzo M, De Rosse A, Ometto L, Mous J, Barelli A, Borri A, Salmaso L, Chieco-Bianchi L (1992) CD4 epitope masking by gp120/anti-gp120 antibody complexes. A potential mechanism for $CD4^+$ cell function down-regulation in AIDS patients. J Immunol 148: 2709–2716

Asjo B, Ivhed I, Gidlund M, Fuerstenberg S, Fenyo EM, Nilsson K, Wigzell H (1987) Susceptibility to infection by the human immunodeficiency virus (HIV) correlates with T4 expression in a parental monocytoid cell line and its subclones. Virology 157: 359–365

Banda NK, Bernier J, Kurahara DK, Kurrie R, Haigwood N, Sekaly R-P, Finkel TH (1992) Cross-linking CD4 by human immunodeficiency virus gp120 primes T cells for activation-induced apoptosis. J Exp Med 176: 1099–1106

Bank I, Chess L (1985) Perturbation of the T4 molecule transmits a negative signal to T cells. J Exp Med 162: 1294–1303

Bergeron L, Sodroski J (1992) Dissociation of unintegrated viral DNA accumulation from single-cell lysis induced by human immunodeficiency virus type 1. J Virol 66: 5777–5787

Boldt-Houle DM, Rinaldo CR Jr, Ehrlich GD (1993) Random depletion of T cells that bear specific T cell receptor V beta sequences in AIDS patients. J Leukoc Biol 54: 486–491

Bolognesi DP, Langlois AJ, Schafer W (1975) Polypeptides of mammalian oncornaviruses IV. Structural components of murine leukemia virus released as soluble antigens in cell culture. Virology 68: 550–555

Boyer V, Smith LR, Ferre F, Pezzoli P, Trauger RJ, Jensen FC, Carlo DJ (1993) T cell receptor V beta repertoire in HIV-infection individuals: lack of evidence for selective V beta deletion. Clin Exp Immunol 92: 437–441

Bozzette SA, McCutchan JA, Spector SA, Wright B, Richman DD (1993) A cross-sectional comparison of persons with syncytium- and non-syncytium-inducing human immunodeficiency virus. J Infect Dis 168: 1374–1379

Brenner BG, Gryllis C, Wainberg MA (1990) Role of antibody-dependent cellular cytotoxicity and lymphokine-activated killer cells in AIDS and related diseases. J Leukoc Biol 50: 628–640

Bullough PA, Hughson FM, Skehel JJ, Wiley DC (1994) Structure of the influenza haemagglutinin at the pH of membrane fusion. Nature 371: 37–43

Callahan KM, Fort MM, Obah EA, Reinherz EL, Siliciano RF (1990) Genetic variability in HIV-1 gp120 affects interactions with HLA molecules and T cell receptor. J Immunol 144: 3341–3346

Callahan KM, Rowell JF, Soloski MJ, Machamer CE, Siliciano RF (1993) HIV-1 envelope protein is expressed on the surface of infected cells before its processing and presentation to class II-restricted T lymphocytes. J Immunol 151: 2928–2942

Chen ZW, Kou ZC, Shen L, Reimann KA, Letvin NL (1993) Conserved T-cell receptor repertoire in simian immunodeficiency virus-infected rhesus monkeys. J Immunol 151: 2177–2187

Cheng-Mayer C, Seto D, Tateno M, Levy JA (1988) Biological features of HIV-1 that correlate with virulence in the host. Science 240: 80–82

Chirmule N, Kalyanaraman V, Oyaizu N, Pahwa S (1988) Inhibitory influences of envelope glycoproteins of HIV-1 on normal immune responses. J Acquir Immune Defic Syndr 1, 425–430

Chirmule N, Kalyanaraman VS, Oyaizu N, Slade HB, Pahwa S (1990) Inhibition of functional properties of tetanus antigen-specific T-cell clones by envelope glycoprotein gp120 of human immunodeficiency virus. Blood 75: 152–159

Clayton LK, Sieh M, Pious DA, Reinherz EL (1989) Identification of human CD4 residues affecting class II MHC versus HIV-1 gp120 binding. Nature 339: 548–551

Crise B, Rose JK (1992) Human immunodeficiency virus type 1 glycoprotein precursor retains a CD4–p56$^{lck}$ complex in the endoplasmic reticulum. J Virol 66: 2296–2301

Crise B, Buonocore L, Rose JK (1990) CD4 is retained in the endoplasmic reticulum by the human immunodeficiency virus type 1 glycoprotein precursor. J Virol 64: 5585–5593

Curiel TJ, Wong JT, Gorczyca PF, Schooley RT, Walker BD (1993) CD4$^{+}$ human immunodeficiency virus type 1 (HIV-1) envelope-specific cytotoxic T lymphocytes derived from the peripheral blood cells of an HIV-1-infected individual. AIDS Res Hum Retroviruses 9: 61–68

Dalgleish AG, Beverley PC, Clapham PR, Crawford DH, Greaves MF, Weiss RA (1984) The CD4 (T4) antigen is an essential component of the receptor for the AIDS retrovirus. Nature 312: 763–767

Daniel V, Susal C, Prodeus AP, Zimmermann R, Huth-Kuhne A, Opelz G (1993a) CD4$^{+}$ lymphocyte depletion in HIV-infected patients is associated with gp120-immunoglobulin-complement attachment to CD4$^{+}$ cells. Vox Sang 64: 31–36

Daniel V, Susal C, Weimer R, Zimmermann R, Huth-Kuhne A, Opelz G (1993b) Association of T cell and macrophage dysfunction with surface gp120-immunoglobulin-complement complexes in HIV-infected patients. Clin Exp Immunol 93: 152–156

Earl PL, Doms RW, Moss B (1990) Oligomeric structure of the human immunodeficiency virus type 1 envelope glycoprotein. Proc Natl Acad Sci USA 87: 648–652

Fenyo EM, Morfeldt-Manson L, Chiodi F, Lind B, von Gegerfelt A, Albert J, Olausson E, Asjo B (1988) Distinctive replicative and cytopathic characteristics of human immunodeficiency virus isolates. J Virol 62: 4414–4419

Fisher AG, Ensoli B, Looney D, Rose A, Gallo RC, Saag MS, Shaw GM, Hahn BH, Wong-Staal F (1988) Biologically diverse molecular variants within a single HIV-1 isolate. Nature 334: 444–447

Gartner S, Markovits P, Markovitz DM, Kaplan MH, Gallo RC, Popovic M (1986) The role of mononuclear phagocytes in HTLV-III/LAV infection. Science 233: 215–219

Gelderblom HR, Reupke H, Pauli G (1985) Loss of envelope antigens of HTLV-III/LAV, a factor in AIDS pathogenesis. Lancet 2: 1016–1017

Gelderblom HR, Hausmann EHS, Ozel M, Pauli G, Koch MA (1987) The structure of human immunodeficiency virus (HIV) and immunolocalization of structural proteins. Virology 156: 171–176
Giorgi JV, Fahey JL, Smith DC, Hultin LE, Cheng E, Mitsuyasu RT, Detels R (1987) Early effects of HIV on CD4⁺ lymphocytes in vivo. J Immunol 138: 3725–3730
Gougeon ML, Garcia S, Heeney J, Tschopp R, Lecoeur H, Guetard D, Rame V, Dauguet C, Montagnier L (1993) Programmed cell death in AIDS-related HIV and SIV infections. AIDS Res Hum Retroviruses 9: 553–563
Groux H, Torpier G, Monte D, Mouton Y, Capron A, Ameisen JC (1992) Activation-induced death by apoptosis in CD4⁺ T cells from human immunodeficiency virus-infected asymptomatic individuals. J Exp Med 175: 331–340
Gurley RJ, Ikeuchi K, Byrn RA, Anderson K, Groopman JE (1989) CD4⁺ lymphocyte function with early human immunodeficiency virus infection. Proc Natl Acad Sci USA 86: 1993–1997
Habeshaw JA, Dalgleish AG, Bountiff L, Newell AL, Wilks D, Walker LC, Manca F (1990) AIDS pathogenesis: HIV envelope and its interaction with cell proteins. Immunol Today 11: 418–425
Hammond SA, Obah E, Stanhope P, Monell CR, Strand M, Robbins FM, Bias WB, Karr RW, Koenig S, Siliciano RF (1991) Characterization of a conserved T cell epitope in HIV-1 gp41 recognized by vaccine-induced human cytolytic T cells. J Immunol 146: 1470–1477
Hammond SA, Bollinger RC, Stanhope PE, Quinn TC, Schwartz D, Clements ML, Siliciano RF (1992) Comparative clonal analysis of human immunodeficiency virus type 1 (HIV-1)-specific CD4⁺ and CD8⁺ cytolytic T lymphocytes isolated from seronegative humans immunized with candidate HIV-1 vaccines. J Exp Med 176: 1531–1542
Harper ME, Marselle LM, Gallo RC, Wong-Staal F (1986) Detection of lymphocytes expressing human T-lymphotropic virus type III in lymph nodes and peripheral blood from infected individuals by *in situ* hybridization. Proc Natl Acad Sci USA 83: 772–776
Hildreth JE, Orentas RJ (1989) Involvement of a leukocyte adhesion receptor (LFA-1) in HIV-induced syncytium formation. Science 244: 1075–1078
Ho DD, Rota TR, Hirsh MS (1986) Infection of monocyte/macrophages by human T lymphotropic virus type III. J Clin Invest 77: 1712–1720
Horak ID, Popovik M, Horak EM, Lucas PJ, Gress RE, June CH, Bolen JB (1990) No T-cell tyrosine protein kinase signaling or calcium mobilization after CD4 association with HIV-1 or HIV-1 gp120. Nature 348: 557–560
Hoxie JA, Alpers JD, Rackowski JL, Huebner K, Haggarty BS, Cedarbaum AJ, Reed JC (1986) Alterations in T4 (CD4) protein and mRNA synthesis in cells infected with HIV. Science 234: 1123–1127
Imberti L, Sottini A, Bettinardi A, Puoti M, Primi D (1991) Selective depletion in HIV infection of T cells that bear specific T cell receptor V beta sequences. Science 254: 860–862
Johnson RP, Hammond SA, Trocha A, Siliciano RF, Walker BD (1994) Induction of a MHC class I-restricted cytotoxic T lymphocyte response to a highly conserved region of HIV-1 gp120 in seronegative humans immunized with a candidate HIV-1 vaccine. J Virol 68: 3145–3153
Klatzmann D, Champagne E, Chamaret S, Gruest J, Guetard D, Hercend T Gluckman JC, Montagnier L (1984) T-lymphocyte T4 molecule behaves as the receptor for the human retrovirus LAV. Nature 312: 767–768
Koenig S, Fauci AS (1988) Immunopathogenesis and immune response to human immunodeficiency virus. In: DeVita V, Hellman S, Rosenberg SA (eds) AIDS. Lippincott, Philadelphia
Koga Y, Sasaki M, Nakamura K, Kimura G, Nomoto K (1990a) Intracellular distribution of the envelope glycoprotein of human immunodeficiency virus and its role in the production of cytopathic effect in CD4⁺ and CD4⁻ human cell lines. J Virol 64: 4661–4671
Koga Y, Sasaki M, Yoshida H, Wigzell H, Kimura G, Nomoto K (1990b) Cytopathic effect determined by the amount of CD4 molecules in human cell lines expressing envelope glycoprotein of HIV. J Immunol 144: 94–102
Koga Y, Sasaki M, Yoshida H, Oh-Tsu M, Kimura G, Nomoto K (1991) Disturbance of nuclear transport of proteins in CD4⁺ cells expressing gp160 of human immunodeficiency virus. J Virol 65: 5609–5612
Koga Y, Nakamura K, Sasaki M, Kimura G, Nomoto K (1994) The difference in gp160 and gp120 of HIV type 1 in the induction of CD4 downregulation preceding single-cell killing. Virology 201: 137–141
Koot M, Keet IP, Vos AH, de Goede RE, Roos MT, Coutinho RA, Miedema F, Schellekens PT, Tersmette M (1993) Prognostic value of HIV-1 syncytium-inducing phenotype for rate of CD4⁺ cell depletion and progression to AIDS. Ann Intern Med 118: 681–688
Kornfeld H, Cruikshank WW, Pyle SW, Berman JS, Center DM (1988) Lymphocyte activation by HIV-1 envelope glycoprotein. Nature 335: 445–448

Kowalski M, Bereron L, Dorfman T, Haseltine W, Sodroski W (1991) Attenuation of human immunodeficiency virus type 1 cytopathic effect by a mutation affecting the transmembrane envelope glycoprotein. J Virol 65: 281–291

Krowka J, Stites D, Mills J, Hollander H, McHugh T, Busch M, Wilhelm L, Blackwood L (1988) Effects of interleukin 2 and large envelope glycoprotein (gp120) of human immunodeficiency virus (HIV) on lymphocyte proliferative responses to cytomegalovirus. Clin Exp Immunol 72: 179–185

Kundu SK, Merigan TC (1992) Equivalent recognition of HIV proteins, Env, Gag and Pol by CD4⁺ and CD8⁺ cytotoxic T-lymphocytes [published erratum appears in AIDS (1992) 6(9): following 1051]. AIDS 6: 643–649

Kundu SK, Katzenstein D, Moses LE, Merigan TC (1992) Enhancement of human immunodeficiency virus (HIV)-specific CD4⁺ and CD8⁺ cytotoxic T-lymphocyte activities in HIV-infected asymptomatic patients given recombinant gp160 vaccine. Proc Natl Acad Sci USA 89: 11204–11208

Landau NR, Warton M, Littman DR (1988) The envelope glycoprotein of the human immunodeficiency virus binds to the immunoglobulin-like domain of CD4. Nature 334: 159–162

Lane HC, Depper JM, Greene WC, Whalen G, Waldmann TA, Fauci (1985) Qualitative analysis of immune function in patients with the acquired immunodeficiency syndrome: evidence for a selective defect in soluble antigen recognition. N Engl J Med 313: 79–84

Lanzavecchia A, Roosnek E, Gregory T, Berman P, Abrignani S (1988) T cells can present antigens such as HIV gp120 targeted to their own surface molecules. Nature 334: 530–532

Lasky LA, Nakamura G, Smith DH, Fennie C, Shimasaki C, Patzer E, Berman P, Gregory T, Capon DJ (1987) Delineation of a region of the human immunodeficiency virus type 1 gp120 glycoprotein critical for interaction with the CD4 receptor. Cell 50: 975–985

Laurence J, Hodtsev AS, Posnett DN (1992) Superantigen implicated in dependence of HIV-1 replication in T cells on TCR V beta expression. Nature 358: 255–259

Laurent-Crawford AG, Hovanessian AG (1993) The cytopathic effect of human immunodeficiency virus is independent of high levels of unintegrated viral DNA accumulated in response to superinfection of cells. J Gen Virol 74: 2619–2628

Laurent-Crawford AG, Krust B, Muller S, Riviere Y, Rey-Cuillė MA, Bechet JM, Montagnier L, Hovanessian AG (1991) The cytopathic effect of HIV is associated with apoptosis. Virology 185: 829–839

Layne SP, Merges MJ, Dembo M, Spouge JL, Conley SR, Moore JP, Raina JL, Renz H, Gelderblom HR, Nara PL (1992) Factors underlying the spontaneous inactivation and susceptibility to neutralization of human immunodeficiency virus. Virology 189: 695–714

Liegler TJ, Stites DP (1994) HIV-1 gp120 and anti-gp120 induce reversible unresponsiveness in peripheral CD4 T lymphocytes. J Acquir Immune Defic Syndr 7: 340–348

Lifson JD, Feinberg MB, Reyes GR, Rabin L, Banapour B, Chakrabarti S, Moss B, Wong-Staal F, Steimer KS, Engleman EG (1986a) Induction of CD4-dependent cell fusion by the HTLV-III/LAV envelope glycoprotein. Nature 323: 725–728

Lifson JD, Reyes GR, McGrath MS, Stein BS, Engleman EG (1986b) AIDS retrovirus induced cytopathology: giant cell formation and involvement of CD4 antigen. Science 232: 1123–1127

Littaua RA, Oldstone MB, Takeda A, Ennis FA (1992) A CD4⁺ cytotoxic T-lymphocyte clone to a conserved epitope on human immunodeficiency virus type 1 p24: cytotoxic activity and secretion of interleukin-2 and interleukin-6. J Virol 66: 608–611

Lu YY, Koga Y, Tanaka K, Sasaki M, Kimura G, Nomoto K (1994) Apoptosis induced in CD4⁺ cells expressing gp160 of human immunodeficiency virus type 1. J Virol 68: 390–399

Lyerly HK, Matthews TJ, Langlois AJ, Bolognesi DP, Weinhold KJ (1987) Human T-cell lymphotropic virus IIIB glycoprotein (gp120) bound to CD4 determinants on normal lymphocytes and expressed by infected cells as a target for immune attack. Proc Natl Acad Sci USA 84: 4601–4605

Macher AM, De Vinatea ML, Angritt P, Tuur SM, Reichert CM (1988) Pathological features of patients infected with human immunodeficiency virus. In: DeVita D, Hellman S, Rosenberg SA (eds) AIDS Lippincott, Philadelphia, pp 155–184

Maddon PJ, Dalgleish AG, McDougal JS, Clapham PR, Weiss RA, Axel R (1986) The T4 gene encodes the AIDS virus receptor and is expressed in the immune system and the brain. Cell 47: 333–348

Manca F, Habeshaw JA, Dalgleish AG (1990) HIV envelope glycoprotein, antigen specific T-cell responses, and soluble CD4. Lancet 335: 811–815

Mann DL, Lasane F, Popovic M, Arthur LO, Robey WG, Blattner WA, Newman MJ (1987) HTLV-III large envelope protein (gp120) suppresses PHA-induced lymphocyte blastogenesis. J Immunol 138: 2640–2644

Martin SJ, Matear PM, Vyakarnam A (1994) HIV-1 infection of human CD4⁺ T cell in vitro. Differential induction of apoptosis in these cells. J Immunol 152: 330–342

McCune JM, Rabin LB, Feinberg MB, Lieberman M, Kosek JC, Reyes GR, Weissman IL (1988) Endoproteolytic cleavage of gp160 is required for the activation of human immunodeficiency virus. Cell 53: 55–67

McDougal JS, Kennedy MS, Sligh JM, Cort SP, Mawle A, Nicholson JK (1986a) Binding of HTLV-III/LAV to T4⁺ T cells by a complex of the 110K viral protein and the T4 molecule. Science 231: 382–385

McDougal JS, Nicholson JK, Cross GD, Cort SP, Kennedy MS, Mawle AC (1986b) Binding of the human retrovirus HTLV-III/LAV/ARV/HIV to the CD4 (T4) molecule: conformation dependence, epitope mapping, antibody inhibition, and potential for idiotypic mimicry. J Immunol 137: 2937–2944

McKeating JA, McKnight A, Moore JP (1991) Differential loss of envelope glycoprotein gp120 from virions of human immunodeficiency virus type 1 isolates: Effects on infectivity and neutralization. J Virol 65: 852–860

Meyaard L, Otto SA, Jonker RR, Mijnster MJ, Keet RPM, Miedema F (1992) Programmed cell death of T cells in HIV-1 infection. Science 257: 217–219

Meyaard L, Otto SA, Keet IP, Roos MT, Miedema F (1994) Programmed death of T cells in human immunodeficiency virus infection. No correlation with progression to disease. J Clin Invest 93: 982–988

Miskovsky EP, Liu AY, Pavlat W, Viveen R, Stanhope PE, Finzi D, Fox WMI, Hruban RH, Podack ER, Siliciano RF (1994) Mechanism of cytolysis by HIV-1-specific CD4⁺ human CTL induced by candidate AIDS vaccines. J Immunol 152: 2787–2799

Mittler RS, Hoffmann MK (1989) Synergism between HIV gp120 and gp120-specific antibody in blocking human T cell activation. Science 245: 1380–1382

Moore JP, McKeating JA, Weiss RA, Sattentau QJ (1990) Dissociation of gp120 from HIV-1 virions induced by soluble CD4. Science 250: 1139–1142

Moore MW, Carbone FW, Bevan MJ (1988) Introduction of soluble antigen into the class I pathway of antigen processing and presentation. Cell 54: 777–786

Morrison LA, Lukacher A, Braciale VL, Fan DP, Braciale TP (1986) Differences in antigen presentation to MHC class I and class II-restricted influenza virus specific cytolytic T lymphocyte clones. J Exp Med 163: 903–913

Newell MK, Haughn LJ, Maroun CR, Julius MH (1990) Death of mature T cells by separate ligation of CD4 and the T-cell antigen receptor. Nature 347: 286–289

Nicloson JKA, Gross GD, Callaway CS, McDougal JS (1986) In vitro infection of human monocytes with T lymphotrophic virus type III/lymphadenopathy-associated virus (HTLV-III/LAV). J Immunol 137: 323–329

Oh S-K, Cruishank WW, Raina J, Blanchard GC, Adler WH, Walker J, Kornfeld H (1992) Identification of HIV-1 envelope glycoproteins in the serum of AIDS and ARC patients. J Acquir Immune Defic Syndr 5: 251–256

Orentas RJ, Hildreth JE, Obah E, Polydefkis M, Smith GE, Clements ML, Siliciano RF (1990) Induction of CD4+ human cytolytic T cells specific for HIV-infected cells by a gp160 subunit vaccine. Science 248: 1234–1237

Polydefkis M, Koenig S, Flexner C, Obah E, Gebo K, Chakrabarti S, Earl PL, Moss B, Siliciano RF (1990) Anchor sequence-dependent endogenous processing of human immunodeficiency virus 1 envelope glycoprotein gp160 for CD4⁺ T cell recognition. J Exp Med 171: 875–887

Posnett DN, Kabak S, Hodtsev AS, Goldberg EA, Asch A (1993) T-cell antigen receptor V beta subsets are not preferentially deleted in AIDS. AIDS 7: 625–631

Richman DD, Bozzette SA (1994) The impact of the syncytium-inducing phenotype of human immunodeficiency virus on disease progression. J Infect Dis 169: 968–974

Rosenstein Y, Burakoff SJ, Herrmann SH (1994) HIV-gp120 can block CD4-class II MHC-mediated adhesion. J Immunol 144: 526–531

Salahuddin SZ, Rose RM, Groopman JE, Markham PD, Gallo RC (1986) Human T lymphotropic virus type III infection of human alveolar macrophages. Blood 68: 281–284

Schneider J, Kaaden O, Copeland TD, Oroszlan S, Hunsmann G (1986) Shedding and interspecies type seo-reactivity of the envelope glycopolypeptide gp120 of the human immunodeficiency virus. J Gen Virol 67: 2533–2538

Schnittman S, Lane HC, Roth J, Burrows A, Folks T, Kehri J, Koenig S, Berman P, Fauci A (1988) Characterization of gp120 binding of CD4 and an assay that measures ability of sera to inhibit this binding. J Immunol 141: 4181–4186

Schwartz S, Felber BK, Fenyo EM, Pavlakis GN (1990) Env and Vpu proteins of human immunodeficiency virus type 1 are produced from multiple bicistronic mRNAs. J Virol 64: 5448–5456

Shalaby MR, Krowka JF, Gregory TJ, Hirabayashi SE, McCabe SM, Kaufman DS, Stites DP, Ammann AJ (1987) The effects of human immunodeficiency virus recombinant envelope glycoprotein on immune cell functions in vitro. Cell Immunol 110: 140–148

Shearer GM, Bernstein DC, Tung KSK, Via CS, Redfield R, Salahuddin SZ, Gallo RC (1986) A model for the selective loss of major histocompatibility complex self-restricted T cell immune responses during the development of acquired immune deficiency syndrome (AIDS). J Immunol 137: 2514–2521

Siliciano RF, Lawton T, Knall C, Karr RC, Berman P, Gregory T, Reinherz EL (1988) Analysis of host-virus interactions in AIDS with anti-gp120 human T cell clones: effect of HIV sequence variation and a mechanism for CD4⁺ cell depletion. Cell 54: 561–575

Siliciano RF, Knall C, Karr RC, Berman P, Gregory T, Reinherz EL (1989) Recognition of human immunodeficiency virus (HIV) glycoprotein gp120 by T cells: role of monocyte CD4 in the presentation of gp120. J Immunol 142: 1506–1511

Sodroski J, Goh WC, Rosen C, Campbell K, Haseltine WA (1986) Role of the HTLV-III/LAV envelope in syncytium formation and cytopathicity. nature 322: 470–474

Somasundaran M, Robinson HL (1987) A major mechanism of human immunodeficiency virus-induced cell killing does not involve cell fusion. J Virol 61: 3114–3119

Stanhope PE, Clements ML, Siliciano RF (1993a) Human CD4⁺ cytolytic T lymphocyte responses to a human immunodeficiency virus type 1 gp160 subunit vaccine. J Infect Dis 168: 92–100

Stanhope PE, Liu AY, Pavlat W, Pitha PM, Clements ML, Siliciano RF (1993b) An HIV-1 envelope protein vaccine elicits a functionally complex human CD4⁺ T cell response that includes cytolytic T lymphocytes. J Immunol 150: 4672–4686

Stevenson M, Meier C, Mann AM, Chapman N, Wasiak A (1988) Envelope glycoprotein of HIV induces interference and cytolysis resistance in CD4⁺ cells: mechanism for persistence in AIDS. Cell 53: 483–496

Stevenson M, Haggerty S, Lamonica C, Mann AM, Meier C, Wasiak A (1990) Cloning and characterization of human immunodeficiency virus type 1 variants diminished in the ability to induce syncytium-independent cytolysis. J Virol 64: 3792–3803

Terai C, Kornbluth RS, Pauza CD, Richman DD, Carson DA (1991) Apoptosis as a mechanism of cell death in cultured T lymphoblasts acutely infected with HIV-1. J Clin Invest 87: 1710–1715

Tersmette M, Gruters RA, de Wolf F, de Goede RE, Lange JM, Schellekens PT, Goudsmit J, Huisman HG, Miedema F (1989) Evidence for a role of virulent human immunodeficiency virus (HIV) variants in the pathogenesis of acquired immunodeficiency syndrome: studies on sequential HIV isolates. J Virol 63: 2118–2125

Thali M, Olshevsky U, Furman C, Gabuzda D, Posner M, Sodroski J (1991) Characterization of a discontinuous human immunodeficiency virus type 1 gp120 epitope recognized by a broadly reactive neutralizing human monoclonal antibody. J Virol 65: 6188–6193

Tyler DS, Nastala CL, Stanley SD, Matthews TJ, Lyerly HK, Bolognesi DP, Weinhold KJ (1989) GP120 specific cellular cytotoxicity in HIV-1 seropositive individuals. Evidence for circulating CD16⁺ effector cells armed in vivo with cytophilic antibody. J Immunol 142: 1177–1182

Tyler DS, Stanley SD, Nastala CA, Austin AA, Bartlett JA, Stine KC, Lyerly HK, Bolognesi DP, Weinhold KJ (1990) Alterations in antibody-dependent cellular cytotoxicity during the course of HIV-1 infection. Humoral and cellular defects. J Immunol 144: 3375–3384

van Noesel CJM, Gruters RA, Terpstra FG, Schellekens PTA, van Lier RAW, Miedema F (1994) Functional and phenotypic evidence for a selective loss of memory T cells in asymptomatic human immunodeficiency virus-infected men. J Clin Invest 86: 293–299

Walker BD, Plata F (1990) Cytotoxic T lymphocytes against HIV (review). AIDS 4: 177–184

Weinhold KJ, Lyerly HK, Stanley SD, Austin AA, Matthews TJ, Bolognesi DP (1989) HIV-1 GP120-mediated immune suppression and lymphocyte destruction in the absence of viral infection. J Immunol 142: 3091–3097

White J, Kielian M, Helenius A (1983) Membrane fusion proteins of enveloped animal viruses. Q Rev Biophys 16: 151–195

Willey RL, Bonifacino JS, Potts BJ, Martin MA, Klausner RD (1988) Biosynthesis, cleavage, and degradation of the human immunodeficiency virus 1 envelope glycoprotein gp160. Proc Natl Acad Sci USA 85: 9580–9584

Willey RL, Maldarelli F, Martin MA, Strebel K (1992a) Human immunodeficiency virus type 1 Vpu protein regulates the formation of intracellular gp160-CD4 complexes. J Virol 66: 226–234

Willey RL, Maldarelli F, Martin MA, Strebel K (1992b) Human immunodeficiency virus type 1 Vpu protein induces rapid degradation of CD4. J Virol 66: 7193–7200

York-Higgins D, Chang-Mayer C, Bauer D, Levy JA, Dina D (1990) Human immunodeficiency virus type 1 cellular host range, replication, and cytopathicity are linked to the envelope region of the viral genome. J Virol 64: 4016–4020

# Subject Index

# Current Topics in Microbiology and Immunology

## Volumes published since 1989 (and still available)

Vol. 162: **Roy, Polly; Gorman, Barry M. (Eds.):** Bluetongue Viruses. 1990. 37 figs. X, 200 pp. ISBN 3-540-51922-X

Vol. 163: **Turner, Peter C.; Moyer, Richard W. (Eds.):** Poxviruses. 1990. 23 figs. X, 210 pp. ISBN 3-540-52430-4

Vol. 164: **Bækkeskov, Steinnun; Hansen, Bruno (Eds.):** Human Diabetes. 1990. 9 figs. X, 198 pp. ISBN 3-540-52652-8

Vol. 165: **Bothwell, Mark (Ed.):** Neuronal Growth Factors. 1991. 14 figs. IX, 173 pp. ISBN 3-540-52654-4

Vol. 166: **Potter, Michael; Melchers, Fritz (Eds.):** Mechanisms in B-Cell Neoplasia. 1990. 143 figs. XIX, 380 pp. ISBN 3-540-52886-5

Vol. 167: **Kaufmann, Stefan H. E. (Ed.):** Heat Shock Proteins and Immune Response. 1991. 18 figs. IX, 214 pp. ISBN 3-540-52857-1

Vol. 168: **Mason, William S.; Seeger, Christoph (Eds.):** Hepadnaviruses. Molecular Biology and Pathogenesis. 1991. 21 figs. X, 206 pp. ISBN 3-540-53060-6

Vol. 169: **Kolakofsky, Daniel (Ed.):** Bunyaviridae. 1991. 34 figs. X, 256 pp. ISBN 3-540-53061-4

Vol. 170: **Compans, Richard W. (Ed.):** Protein Traffic in Eukaryotic Cells. Selected Reviews. 1991. 14 figs. X, 186 pp. ISBN 3-540-53631-0

Vol. 171: **Kung, Hsing-Jien; Vogt, Peter K. (Eds.):** Retroviral Insertion and Oncogene Activation. 1991. 18 figs. X, 179 pp. ISBN 3-540-53857-7

Vol. 172: **Chesebro, Bruce W. (Ed.):** Transmissible Spongiform Encephalopathies. 1991. 48 figs. X, 288 pp. ISBN 3-540-53883-6

Vol. 173: **Pfeffer, Klaus; Heeg, Klaus; Wagner, Hermann; Riethmüller, Gert (Eds.):** Function and Specificity of γ/δ TCells. 1991. 41 figs. XII, 296 pp. ISBN 3-540-53781-3

Vol. 174: **Fleischer, Bernhard; Sjögren, Hans Olov (Eds.):** Superantigens. 1991. 13 figs. IX, 137 pp. ISBN 3-540-54205-1

Vol. 175: **Aktories, Klaus (Ed.):** ADP-Ribosylating Toxins. 1992. 23 figs. IX, 148 pp. ISBN 3-540-54598-0

Vol. 176: **Holland, John J. (Ed.):** Genetic Diversity of RNA Viruses. 1992. 34 figs. IX, 226 pp. ISBN 3-540-54652-9

Vol. 177: **Müller-Sieburg, Christa; Torok-Storb, Beverly; Visser, Jan; Storb, Rainer (Eds.):** Hematopoietic Stem Cells. 1992. 18 figs. XIII, 143 pp. ISBN 3-540-54531-X

Vol. 178: **Parker, Charles J. (Ed.):** Membrane Defenses Against Attack by Complement and Perforins. 1992. 26 figs. VIII, 188 pp. ISBN 3-540-54653-7

Vol. 179: **Rouse, Barry T. (Ed.):** Herpes Simplex Virus. 1992. 9 figs. X, 180 pp. ISBN 3-540-55066-6

Vol. 180: **Sansonetti, P. J. (Ed.):** Pathogenesis of Shigellosis. 1992. 15 figs. X, 143 pp. ISBN 3-540-55058-5

Vol. 181: **Russell, Stephen W.; Gordon, Siamon (Eds.):** Macrophage Biology and Activation. 1992. 42 figs. IX, 299 pp. ISBN 3-540-55293-6

Vol. 182: **Potter, Michael; Melchers, Fritz (Eds.):** Mechanisms in B-Cell Neoplasia. 1992. 188 figs. XX, 499 pp. ISBN 3-540-55658-3

Vol. 183: **Dimmock, Nigel J.:** Neutralization of Animal Viruses. 1993. 10 figs. VII, 149 pp. ISBN 3-540-56030-0

Vol. 184: **Dunon, Dominique; Mackay, Charles R.; Imhof, Beat A. (Eds.):** Adhesion in Leukocyte Homing and Differentiation. 1993. 37 figs. IX, 260 pp. ISBN 3-540-56756-9

Vol. 185: **Ramig, Robert F. (Ed.):** Rotaviruses. 1994. 37 figs. X, 380 pp. ISBN 3-540-56761-5

Vol. 186: **zur Hausen, Harald (Ed.):** Human Pathogenic Papillomaviruses. 1994. 37 figs. XIII, 274 pp. ISBN 3-540-57193-0

Vol. 187: **Rupprecht, Charles E.; Dietzschold, Bernhard; Koprowski, Hilary (Eds.):** Lyssaviruses. 1994. 50 figs. IX, 352 pp. ISBN 3-540-57194-9

Vol. 188: **Letvin, Norman L.; Desrosiers, Ronald C. (Eds.):** Simian Immunodeficiency Virus. 1994. 37 figs. X, 240 pp. ISBN 3-540-57274-0

Vol. 189: **Oldstone, Michael B. A. (Ed.):** Cytotoxic T-Lymphocytes in Human Viral and Malaria Infections. 1994. 37 figs. IX, 210 pp. ISBN 3-540-57259-7

Vol. 190: **Koprowski, Hilary; Lipkin, W. Ian (Eds.):** Borna Disease. 1995. 33 figs. IX, 134 pp. ISBN 3-540-57388-7

Vol. 191: **ter Meulen, Volker; Billeter, Martin A. (Eds.):** Measles Virus. 1995. 23 figs. IX, 196 pp. ISBN 3-540-57389-5

Vol. 192: **Dangl, Jeffrey L. (Ed.):** Bacterial Pathogenesis of Plants and Animals. 1994. 41 figs. IX, 343 pp. ISBN 3-540-57391-7

Vol. 193: **Chen, Irvin S. Y.; Koprowski, Hilary; Srinivasan, Alagarsamy; Vogt, Peter K. (Eds.):** Transacting Functions of Human Retroviruses. 1995. 49 figs. IX, 240 pp. ISBN 3-540-57901-X

Vol. 194: **Potter, Michael; Melchers, Fritz (Eds.):** Mechanisms in B-cell Neoplasia. 1995. 152 figs. XXV, 458 pp. ISBN 3-540-58447-1

Vol. 195: **Montecucco, Cesare (Ed.):** Clostridial Neurotoxins. 1995. 28 figs. XI., 278 pp. ISBN 3-540-58452-8

Vol. 196: **Koprowski, Hilary; Maeda, Hiroshi (Eds.):** The Role of Nitric Oxide in Physiology and Pathophysiology. 1995. 21 figs. IX, 90 pp. ISBN 3-540-58214-2

Vol. 197: **Meyer, Peter (Ed.):** Gene Silencing in Higher Plants and Related Phenomena in Other Eukaryotes. 1995. 17 figs. IX, 232 pp. ISBN 3-540-58236-3

Vol. 198: **Griffiths, Gillian M.; Tschopp, Jürg (Eds.):** Pathways for Cytolysis. 1995. 45 figs. IX, 224 pp. ISBN 3-540-58725-X

Vol. 199/I: **Doerfler, Walter; Böhm, Petra (Eds.):** The Molecular Repertoire of Adenoviruses I. 1995. 51 figs. XIII, 280 pp. ISBN 3-540-58828-0

Vol. 199/II: **Doerfler, Walter; Böhm, Petra (Eds.):** The Molecular Repertoire of Adenoviruses II. 1995. 36 figs. XIII, 278 pp. ISBN 3-540-58829-9

Vol. 199/III: **Doerfler, Walter; Böhm, Petra (Eds.):** The Molecular Repertoire of Adenoviruses III. 1995. 51 figs. XIII, 310 pp. ISBN 3-540-58987-2

Vol. 200: **Kroemer, Guido; Martinez-A., Carlos (Eds.):** Apoptosis in Immunology. 1995. 14 figs. XI, 242 pp. ISBN 3-540-58756-X

Vol. 201: **Kosco-Vilbois, Marie H. (Ed.):** An Antigen Depository of the Immune System: Follicular Dendritic Cells. 1995. 39 figs. IX, 209 pp. ISBN 3-540-59013-7

Vol. 202: **Oldstone, Michael B. A.; Vitković, Ljubiša (Eds.):** HIV and Dementia. 1995. 40 figs. XIII, 279 pp. ISBN 3-540-59117-6

Vol. 203: **Sarnow, Peter (Ed.):** Cap-Independent Translation. 1995. 31 figs. XI, 183 pp. ISBN 3-540-59121-4

Vol. 204: **Saedler, Heinz; Gierl, Alfons (Eds.):** Transposable Elements. 1995. 42 figs. IX, 234 pp. ISBN 3-540-59342-X

# Springer-Verlag and the Environment

We at Springer-Verlag firmly believe that an international science publisher has a special obligation to the environment, and our corporate policies consistently reflect this conviction.

We also expect our business partners – paper mills, printers, packaging manufacturers, etc. – to commit themselves to using environmentally friendly materials and production processes.

The paper in this book is made from low- or no-chlorine pulp and is acid free, in conformance with international standards for paper permanency.

GPSR Compliance

*The European Union's (EU) General Product Safety Regulation (GPSR) is a set of rules that requires consumer products to be safe and our obligations to ensure this.*

*If you have any concerns about our products, you can contact us on ProductSafety@springernature.com*

In case Publisher is established outside the EU, the EU authorized representative is:

Springer Nature Customer Service Center GmbH
Europaplatz 3
69115 Heidelberg, Germany

**Batch number: 10620141**

Printed by Printforce, the Netherlands